DEFENSE, SECURITY AND STRATEGIES

CYBERSECURITY AND RELATED FEDERAL LAWS

REVISION PROPOSALS

DEFENSE, SECURITY AND STRATEGIES

Additional books in this series can be found on Nova's website under the Series tab.

Additional E-books in this series can be found on Nova's website under the E-book tab.

INTERNET POLICIES AND ISSUES

Additional books in this series can be found on Nova's website under the Series tab.

Additional E-books in this series can be found on Nova's website under the E-book tab.

DEFENSE, SECURITY AND STRATEGIES

CYBERSECURITY AND RELATED FEDERAL LAWS

REVISION PROPOSALS

JAMES R. HENDERSON
AND
DAVID I. COX
EDITORS

Nova Science Publishers, Inc.
New York

For permission to use material from this book please contact us:
Telephone 631-231-7269; Fax 631-231-8175
Web Site: http://www.novapublishers.com

Additional color graphics may be available in the e-book version of this book.

Library of Congress Cataloging-in-Publication Data

ISBN 978-1-62257-743-9

Published by Nova Science Publishers, Inc. † New York

CONTENTS

Preface		**vii**
Chapter 1	Federal Laws Relating to Cybersecurity: Discussion of Proposed Revisions *Eric A. Fischer*	**1**
Chapter 2	Cybersecurity: Selected Legal Issues *Edward C. Liu, Gina Stevens, Kathleen Ann Ruane, Alissa M. Dolan and Richard M. Thompson II*	**81**
Index		**143**

PREFACE

For more than a decade, various experts have expressed increasing concerns about cybersecurity, in light of the growing frequency, impact, and sophistication of attacks on information systems in the United States and abroad. Consensus has also been building that the current legislative framework for cybersecurity might need to be revised. This book discusses cybersecurity and proposals to amend related federal laws, while also examining relevant legal issues.

Chapter 1 – For more than a decade, various experts have expressed increasing concerns about cybersecurity, in light of the growing frequency, impact, and sophistication of attacks on information systems in the United States and abroad. Consensus has also been building that the current legislative framework for cybersecurity might need to be revised.

The complex federal role in cybersecurity involves both securing federal systems and assisting in protecting nonfederal systems. Under current law, all federal agencies have cybersecurity responsibilities relating to their own systems, and many have sector-specific responsibilities for critical infrastructure.

More than 50 statutes address various aspects of cybersecurity either directly or indirectly, but there is no overarching framework legislation in place. While revisions to most of those laws have been proposed over the past few years, no major cybersecurity legislation has been enacted since 2002.

Chapter 2 – The federal government's role in protecting U.S. citizens and critical infrastructure from cyber attacks has been the subject of recent congressional interest. Critical infrastructure commonly refers to those entities that are so vital that their incapacitation or destruction would have a debilitating impact on national security, economic security, or the public health and safety. This report discusses selected legal issues that frequently arise in the context of recent legislation to address vulnerabilities of critical infrastructure to cyber threats, efforts to protect government networks from cyber threats, and proposals to facilitate and encourage sharing of cyber threat information among private sector and government entities. This report also discusses the degree to which federal law may preempt state law.

It has been argued that, in order to ensure the continuity of critical infrastructure and the larger economy, a regulatory framework for selected critical infrastructure should be created to require a minimum level of security from cyber threats. On the other hand, others have argued that such regulatory schemes would not improve cybersecurity while increasing the costs to businesses, expose businesses to additional liability if they fail to meet the imposed cybersecurity standards, and increase the risk that proprietary or confidential business information may be inappropriately disclosed.

In order to protect federal information networks, the Department of Homeland Security (DHS), in conjunction with the National Security Agency (NSA), uses a network intrusion system that monitors all federal agency networks for potential attacks. Known as EINSTEIN, this system raises significant privacy implications—a concern acknowledged by DHS, interest groups, academia, and the general public. DHS has developed a set of procedures to address these concerns such as minimization of information collection, training and accountability requirements, and retention rules. Notwithstanding these steps, there are concerns that the program may implicate privacy interests protected under the Fourth Amendment.

Although many have argued that there is a need for federal and state governments, and owners and operators of the nation's critical infrastructures, to share information on cyber vulnerabilities and threats, obstacles to information sharing may exist in current laws protecting electronic communications or in antitrust law. Private entities that share information may also be concerned that sharing or receiving such information may lead to increased civil liability, or that shared information may contain proprietary or confidential business information that may be used by competitors or government regulators for unauthorized purposes.

Several bills in the 112th Congress would seek to improve the nation's cybersecurity, and may raise some or all of the legal issues mentioned above. For example, H.R. 3523 (Rogers (Mich.)- Ruppersberger) addresses information sharing between the intelligence community and the private sector. H.R. 3674 (Lungren) includes provisions regarding the protection of critical infrastructure, as well as information sharing. H.R. 4257 (Issa-

Cummings) would require all federal agencies to continuously monitor their computer networks for malicious activity and would impose additional cybersecurity requirements on all federal agencies. S. 2102 (Feinstein) seeks to facilitate information sharing. S. 2105 (Lieberman) includes the information sharing provisions of S. 2102, as well as provisions relating to the protection of critical infrastructure and federal government networks. S. 2151 (McCain) and H.R. 4263 (Bono-Mack) also addresses information sharing among the private sector and between the private sector and the government. Many of these bills also include provisions specifically addressing the preemption of state laws.

In: Cybersecurity and Related Federal Laws ISBN 978-1-62257-743-9
Editors: J. R. Henderson and D. I. Cox

Chapter 1

FEDERAL LAWS RELATING TO CYBERSECURITY: DISCUSSION OF PROPOSED REVISIONS*

Eric A. Fischer

SUMMARY

For more than a decade, various experts have expressed increasing concerns about cybersecurity, in light of the growing frequency, impact, and sophistication of attacks on information systems in the United States and abroad. Consensus has also been building that the current legislative framework for cybersecurity might need to be revised.

The complex federal role in cybersecurity involves both securing federal systems and assisting in protecting nonfederal systems. Under current law, all federal agencies have cybersecurity responsibilities relating to their own systems, and many have sector-specific responsibilities for critical infrastructure.

More than 50 statutes address various aspects of cybersecurity either directly or indirectly, but there is no overarching framework legislation in place. While revisions to most of those laws have been proposed over the past few years, no major cybersecurity legislation has been enacted since 2002.

---

* This is an edited, reformatted and augmented version of a Congressional Research Service publication, CRS Report for Congress R42114, prepared for Members and Committees of Congress, from www.crs.gov, dated June 29, 2012.

Recent legislative proposals, including many bills introduced in the 111th and 112th Congresses, have focused largely on issues in 10 broad areas (see "Selected Issues Addressed in Proposed Legislation" for an overview of how current legislative proposals would address issues in several of those areas):

- national strategy and the role of government,
- reform of the Federal Information Security Management Act (FISMA),
- protection of critical infrastructure (including the electricity grid and the chemical industry),
- information sharing and cross-sector coordination,
- breaches resulting in theft or exposure of personal data such as financial information,
- cybercrime,
- privacy in the context of electronic commerce,
- international efforts,
- research and development, and
- the cybersecurity workforce.

For most of those topics, at least some of the bills addressing them have proposed changes to current laws. Several of the bills specifically focused on cybersecurity have received committee or floor action, but none have become law.

Comprehensive legislative proposals on cybersecurity that have received considerable attention in 2012 are S. 2105, recommendations from a House Republican task force, and a proposal by the Obama Administration. They differ in approach, with S. 2105 proposing the most extensive regulatory framework and organizational changes of the three, and the task force recommendations focusing more on incentives for improving private-sector cybersecurity. An alternative to S. 2105, S. 3342 (a refinement of S. 2151), does not include enhanced regulatory authority or new federal entities, but does include cybercrime provisions.

Several narrower House bills have been introduced that address some of the issues raised and recommendations made by the House task force. Four passed the House the week of April 23:

- Cybersecurity Enhancement Act of 2011 (H.R. 2096), which addresses federal cybersecurity R&D and the development of technical standards;
- Cyber Intelligence Sharing and Protection Act (H.R. 3523), which focuses on information sharing and coordination, including sharing of classified information;

- Advancing America's Networking and Information Technology Research and Development Act of 2012 (H.R. 3834), which addresses R&D in networking and information technology, including but not limited to security; and
- Federal Information Security Amendments Act of 2012 (H.R. 4257), which addresses FISMA reform.

One was ordered reported out of the full committee but did not come to the floor:

- Promoting and Enhancing Cybersecurity and Information Sharing Effectiveness Act of 2011 or PRECISE Act of 2011 (H.R. 3674), which addresses the role of the Department of Homeland Security in cybersecurity, including protection of federal systems, personnel, R&D, information sharing, and public/private sector collaboration in protecting critical infrastructure;

Together, those House and Senate bills address most of the issues listed above, although in different ways. All include proposed revisions to some existing laws covered in this report.

INTRODUCTION

For more than a decade, various experts have expressed concerns about information-system security—often referred to as *cybersecurity*—in the United States and abroad.[1] The frequency, impact, and sophistication of attacks on those systems has added urgency to the concerns.[2] Consensus has also been growing that the current legislative framework for cybersecurity might need to be revised to address needs for improved cybersecurity, especially given the continuing evolution of the technology and threat environments. This report discusses that framework and proposals to amend more than 30 acts of Congress that are part of or relevant to it.

Current Legislative Framework

The federal role in addressing cybersecurity is complex. It involves both securing federal systems and fulfilling the appropriate federal role in protecting nonfederal systems. There is as yet no overarching framework legislation in place, but many enacted statutes address various aspects of cybersecurity. Some notable provisions are in the following acts:

- *The Counterfeit Access Device and Computer Fraud and Abuse Act of 1984* prohibits various attacks on federal computer systems and on those used by banks and in interstate and foreign commerce.
- *The Electronic Communications Privacy Act of 1986 (ECPA)* prohibits unauthorized electronic eavesdropping.
- *The Computer Security Act of 1987* gave the National Institute of Standards and Technology (NIST) responsibility for developing security standards for federal computer systems, except the national security systems[3] that are used for defense and intelligence missions, and gave responsibility to the Secretary of Commerce for promulgating security standards.
- *The Paperwork Reduction Act of 1995* gave the Office of Management and Budget (OMB) responsibility for developing cybersecurity policies.
- *The Clinger-Cohen Act of 1996* made agency heads responsible for ensuring the adequacy of agency information-security policies and procedures, established the chief information officer (CIO) position in agencies, and gave the Secretary of Commerce authority to make promulgated security standards mandatory.
- *The Homeland Security Act of 2002 (HSA)* gave the Department of Homeland Security (DHS) some cybersecurity responsibilities in addition to those implied by its general responsibilities for homeland security and critical infrastructure.
- *The Cyber Security Research and Development Act,* also enacted in 2002, established research responsibilities in cybersecurity for the National Science Foundation (NSF) and NIST.
- *The E-Government Act of 2002* serves as the primary legislative vehicle to guide federal IT management and initiatives to make information and services available online, and includes various cybersecurity requirements.
- *The Federal Information Security Management Act of 2002 (FISMA)* clarified and strengthened NIST and agency cybersecurity responsibilities, established a central federal incident center, and made OMB, rather than the Secretary of Commerce, responsible for promulgating federal cybersecurity standards.

More than 40 other laws identified by CRS also have provisions relating to cybersecurity (see *Table 2*). Revisions to many of those laws have been proposed. More than 40 bills and resolutions with provisions related to cybersecurity have been introduced in the 112th Congress, including several proposing revisions to current laws. In the 111th Congress, the total was more than 60.4 Several bills in both Congresses received committee or floor action, but none have become law. In fact, no comprehensive cybersecurity legislation has been enacted since 2002.[5]

Executive Branch Actions

Some significant executive actions have been taken, however.[6] The George W. Bush Administration established the Comprehensive National Cybersecurity Initiative (CNCI) in 2008 through National Security Presidential Directive 54 / Homeland Security Presidential Directive 23 (NSPD-54/HSPD-23). Those documents are classified, but the Obama Administration released a description of them in March 2010.[7] Goals of the 12 subinitiatives in that description include consolidating external access points to federal systems; deploying intrusion detection and prevention systems across those systems; improving research coordination and prioritization and developing "next-generation" technology, information sharing, and cybersecurity education and awareness; mitigating risks from the global supply chain for information technology; and clarifying the federal role in protecting critical infrastructure.

In December 2009, the Obama Administration appointed Howard Schmidt to the position of White House Cybersecurity Coordinator.[8] He is a member of the White House national security staff and is responsible for government-wide coordination of cybersecurity, including the CNCI. One of the most visible initiatives in which he has been involved is the implementation of automated, continuous monitoring of federal information systems.[9] Other stated priorities include developing a unified strategy for network security and incident response, and strengthening partnerships with the private sector and other countries. He works with both the National Security and Economic Councils in the White House. However, the position has no direct control over agency budgets, and some observers argue that operational entities such as the National Security Agency (NSA) have far greater influence and authority.[10] The Obama Administration has also launched several initiatives.[11]

Under current law, all federal agencies have cybersecurity responsibilities relating to their own systems, and many have sector-specific responsibilities for critical infrastructure, such as the Department of Transportation for the transportation sector.

Cross-agency responsibilities are complex, and any brief description is necessarily oversimplified. In general, in addition to the roles of White House entities, DHS is the primary civil-sector cybersecurity agency. NIST, in the Department of Commerce, develops cybersecurity standards and guidelines that are promulgated by OMB, and the Department of Justice is largely responsible for the enforcement of laws relating to cybersecurity.[12] The National Science Foundation (NSF), NIST, and DHS all perform research and development (R&D) related to cybersecurity. The National Security Agency (NSA) is the primary cybersecurity agency in the national security sector, although other agencies also play significant roles. The recently established U.S. Cyber Command, part of the U.S. Strategic Command in the Department of Defense (DOD), has primary responsibility for military cyberspace operations.

Legislative Proposals

In general, legislative proposals on cybersecurity in the 111th and 112th Congresses have focused largely on issues in 10 broad areas:

- national strategy and the role of government,
- reform of FISMA,
- protection of critical infrastructure (especially the electricity grid and the chemical industry),
- information sharing and cross-sector coordination,
- breaches resulting in theft or exposure of personal data such as financial information,
- cybercrime offenses and penalties,
- privacy in the context of electronic commerce,
- international efforts,
- research and development (R&D), and
- the cybersecurity workforce.

For most of those topics, at least some of the bills addressing them proposed changes to current laws.[13]

Selected Legislative Proposals in the 112th Congress

There appears to be considerable support in principle for significant legislation to address most of those issues. The House, Senate, and White House have taken somewhat different approaches to such legislation.

The Senate has been working since last year on a comprehensive bill synthesizing approaches proposed by the Homeland Security and Governmental Affairs Committee (S. 3480 in the 111th Congress and S. 413 in the 112th), the Commerce, Science, and Transportation Committee (S. 773 in the 111th Congress), and others. S. 2105, the Cybersecurity Act of 2012, which includes features of both those bills and others,[14] was introduced in February 2012. An alternative Senate bill, S. 3342, the SECURE IT Act,[15] is a revision of S. 2151, which was originally introduced in March.[16] Several other Senate bills would address specific aspects of cybersecurity, such as data breaches of personal information and cybercrime.

In April 2011, the White House sent a comprehensive, seven-part legislative proposal (*White House Proposal*) to Congress.[17] Some elements of that proposal have been included in both House and Senate bills.

In October, the 12-Member House Republican Cybersecurity Task Force, which had been formed by Speaker Boehner in June, released a series of recommendations (*Task Force Report*) to be used by House committees in developing cybersecurity legislation.[18] Unlike the other proposals, it was not presented in the form of a bill or bills. Several House bills have been introduced subsequently that address some of the issues raised and recommendations made by the *Task Force Report.* Four passed the House the week of April 23:

- Cybersecurity Enhancement Act of 2011 (H.R. 2096), which would addresses federal cybersecurity R&D and the development of technical standards;
- Cyber Intelligence Sharing and Protection Act (H.R. 3523), which focuses on information sharing and coordination, including sharing of classified information;[19]
- Advancing America's Networking and Information Technology Research and Development Act of 2012 (H.R. 3834), which addresses R&D in networking and information technology, including but not limited to security;[20] and
- Federal Information Security Amendments Act of 2012 (H.R. 4257), which addresses FISMA reform.

A fifth bill was ordered reported out of full committee on April 18 but was not included in the cybersecurity bills debated on the House floor the week of April 23:[21]

- Promoting and Enhancing Cybersecurity and Information Sharing Effectiveness Act of 2011 or PRECISE Act of 2011 (H.R. 3674), which addresses the role of the Department of Homeland Security in cybersecurity, including protection of federal systems, personnel, R&D, information sharing, and public/private sector collaboration in protecting critical infrastructure.

Specific issues addressed by several of those bills and proposals are noted in *Table 1*. Together, they address most of the issues listed above, although in different ways. All include or discuss proposed revisions to some existing laws covered in this report.

Those addressed in the House bills are

- "Cyber Security Research and Development Act, 2002" (H.R. 2096, S. 2105, S. 2151, S. 3342);
- "Federal Information Security Management Act of 2002 (FISMA)" (H.R. 4257, the *Task Force Report*, S. 2105, S. 2151, S. 3342, the *White House Proposal*);
- "High Performance Computing Act of 1991" (H.R. 3834, S. 2105, S. 2151, S. 3342)
- "Homeland Security Act of 2002 (HSA)" (H.R. 3674, S. 2105, the *White House Proposal*); and
- "National Security Act of 1947" (H.R. 3523).

Those addressed in other legislative proposals are

- "Antitrust Laws and Section 5 of the Federal Trade Commission Act" (*Task Force Report*, S. 2151, S. 3342)
- "Clinger-Cohen Act (Information Technology Management Reform Act) of 1996" (S. 2105, *White House Proposal*);[22]
- "Counterfeit Access Device and Computer Fraud and Abuse Act of 1984" (*Task Force Report*, S. 2151, S. 3342, *White House Proposal*);
- "E-Government Act of 2002" (*White House Proposal*);

Table 1. Comparison of Topics Addressed by Selected Legislative Proposals on Cybersecurity in the 112th Congress

Topic	Selected House Bills	Task Force Report	S. 2105	S. 3342 (S. 2151)	White House Proposal
DHS authorities for protection of federal systems	H.R. 3674	X	X		X
New DHS office/center	H.R. 3674		X		X
Cybersecurity workforce authorities and programs	H.R. 2096 H.R. 3674	X	X	X	X
	H.R. 3834				
Supply-chain vulnerabilities	H.R. 3674	X	X		X
Cybersecurity R&D	H.R. 2096	X	X	X	X
	H.R. 3674				
	H.R. 3834				
FISMA reform	H.R. 4257	X	X	X	X
Protection of privately held critical infrastructure (CI)	H.R. 3674	X	X		X
Government/private-sector collaboration on CI protection	H.R. 3674	X	X		X
Additional regulation of privately held critical infrastructure		X	X		X
Information sharing	H.R. 3523	X	X	X	X
	H.R. 3674				
FOIA exemption for cybersecurity information	H.R. 3523	X	X	X	X
New information-sharing entities	(H.R. 3674)[a]	X	X		
Public awareness	H.R. 2096	X	X		X
Cybercrime law		X		X	X
Data breach notification		X			X
Internet security provider code of conduct		X			
National security/defense and federal civil sector coordination		X			

Source: CRS.

Note: S. 3342 is a revised version of S. 2151.

[a] The subcommittee version of this bill would have created a new nonprofit quasi-governmental information-sharing entity, but the committee version omitted those provisions (see "Information Sharing").

- "Electronic Communications Privacy Act of 1986 (ECPA)" (*Task Force Report*);
- "Identity Theft Penalty Enhancement Act" (*Task Force Report*); and
- "Racketeer Influenced and Corrupt Organizations Act (RICO)" (*Task Force Report*).

Also, some legislative proposals would provide exemptions under the "Freedom of Information Act (FOIA)" for certain kinds of information provided to the federal government (*Task Force Report,* H.R. 3523, S. 2105, S. 2151, S. 3342, *White House Proposal*). H.R. 3523, S. 2151, and S. 3342 would also permit information sharing that might otherwise be subject to antitrust or other restrictions on sharing,[23] and the *Task Force Report* states that an antitrust exemption might be necessary.

Selected Issues Addressed in Proposed Legislation

The proposals listed in *Table 1* take a range of approaches to address issues in cybersecurity. The discussion below compares those approaches for several issues—"DHS Authorities for Protection of Federal Systems," the "Cybersecurity Workforce," "Research and Development," "FISMA Reform," "Protection of Privately Held Critical Infrastructure (CI)," and "Information Sharing.

DHS Authorities for Protection of Federal Systems

DHS currently has very limited statutory responsibility for the protection of federal information systems. The degree to which its role should be modified has been a matter of some debate. Five of the legislative proposals listed in *Table 1* address DHS authorities for federal civil systems.[24] All five bills would enhance DHS authorities, although to varying degrees and in varying ways.

The *Task Force Report* proposes that Congress "formalize" DHS's current coordinating role in cybersecurity. H.R. 3674 would add new provisions on DHS cybersecurity activities to Title II of HSA; S. 2105 and the *White House Proposal* would add a new subtitle to HSA. All three proposals would provide specific authorities and responsibilities to DHS for risk assessments, protective capabilities, and operational cybersecurity activities.

S. 2105 would also create a new, consolidated DHS cybersecurity and communications center with a Senate-confirmed director who would be responsible for managing federal cybersecurity efforts; for developing and implementing information-security policies, principles, and guidelines; and other functions, including risk assessments and other activities to protect federal systems. The *White House Proposal* would provide such enhanced authority to the DHS Secretary rather than a new center. However, the *White House Proposal* would require the Secretary to establish a center with responsibilities for protecting federal information systems, facilitating information sharing, and coordinating incident response. H.R. 3674 would establish a DHS center with responsibility for information sharing (see "Information Sharing") and technical assistance, and would authorize DHS to conduct specific activities to protect federal systems, including risk assessments and access to agency information-system traffic.

S. 2151 would not amend the HSA but would provide the Secretary of Homeland Security with new responsibilities under FISMA. S. 3342 omits some of those responsibilities and modifies others (see "FISMA Reform").

Cybersecurity Workforce

Concerns have been raised for several years about the size, skills, and preparation of the federal and private-sector cybersecurity workforce.[25] Six proposals in *Table 1* would address those concerns in various ways:

- Provide additional federal hiring and compensation authorities (*Task Force Report*, H.R. 3674, S. 2105, *White House Proposal*).
- Establish or enhance educational programs for development of next-generation cybersecurity professionals[26] (*Task Force Report*, H.R. 2096, H.R. 3834, S. 2105, S. 2151, S. 3342).
- Assess workforce needs (H.R. 2096, S. 2105, S. 2151, S. 3342).
- Use public/private-sector personnel exchanges (*Task Force Report*, *White House Proposal*).

Research and Development

The need for improvements in fundamental knowledge of cybersecurity and new solutions and approaches has been recognized for well over a decade[27] and was a factor in the passage of the Cybersecurity Research and Development Act in 2002 (P.L. 107-305, H.Rept. 107-355). That law focuses on cybersecurity R&D by NSF and NIST. The Homeland Security Act of 2002, in contrast, does not specifically mention cybersecurity R&D. However, DHS and several other agencies make significant investments in it. About 60% of reported funding by agencies in cybersecurity and information assurance is defense-related (invested by the Defense Advanced Research Projects Agency [DARPA], NSA, and other defense agencies), with NSF accounting for about

15%, NIST, DHS, and DOE 5%-10% each.[28] Seven of the nine legislative proposals in *Table 1* address cybersecurity R&D. Five would establish requirements for R&D on specific topics such as detection of threats and intrusions, identity management, test beds, and supply-chain security. Agencies for which the proposals include provisions specifying research topics or providing funding authorization include

- DHS (H.R. 3674, S. 2105),
- NIST (H.R. 2096, S. 2151, S. 3342),
- NSF (H.R. 2096, S. 2105, S. 2151, S. 3342), and
- Multiagency[29] (H.R. 3834, S. 2105, S. 2151, S. 3342).

The *Task Force Report*, H.R. 2096, H.R. 3834, S. 2105, S. 2151, and S. 3342 address planning and coordination of research among federal agencies through the White House National Science and Technology Council (NSTC) and other entities.

The *White House Proposal* does not include any specific R&D provisions but includes cybersecurity R&D among a set of proposed requirements for the Secretary of Homeland Security.

FISMA Reform

The "Federal Information Security Management Act of 2002 (FISMA)" was enacted in 2002. It revised the framework that had been enacted in several previous laws (see *Table 2*).

FISMA has been criticized for focus on procedure and reporting rather than operational security, a lack of widely accepted cybersecurity metrics, variations in agency interpretation of the mandates in the act, excessive focus on individual information systems as opposed to the agency's overall information architecture, and insufficient means to enforce compliance both within and across agencies. Five legislative proposals in the 112th Congress (the *Task Force Report*, H.R. 4257, S. 2105, S. 2151, S. 3342, and the *White House Proposal*) would revise FISMA, while retaining much of the current framework:

- All five would continue requirements for agency-wide information security programs, annual independent review of security programs, and reports on program effectiveness and deficiencies.

- All include requirements for continuous monitoring of agency systems, including automated monitoring.
- All would retain the responsibility of NIST for development of cybersecurity standards, including compulsory standards. H.R. 4257 would retain OMB's current responsibility for promulgating the standards, whereas S. 2105, S. 2151, S. 3342, and the *White House Proposal* would transfer that responsibility to the Secretary of Commerce. [30]
- H.R. 4257 would also retain OMB's current responsibility for overseeing federal information-security policy and evaluating agency information-security programs. S. 2105 and the *White House Proposal* would transfer authorities and functions for information security policy from OMB to DHS. OMB has already delegated some authorities to DHS administratively,[31] and the *Task Force Report* expresses support for that approach. S. 2151 and S. 3342, in contrast, would transfer that responsibility to the Secretary of Commerce. However, none of the proposals would give the Secretaries of Commerce or Homeland Security authority to approve or disapprove agency information security plans. Only H.R. 4257 would expressly retain OMB's current power to use its financial authority to enforce accountability.
- S. 2105 and the *White House Proposal* would provide new protective authorities to the Secretary of Homeland Security, including intrusion detection, use of countermeasures, access to communications and other system traffic at agencies, as well as the power to direct agencies to take protective actions and, in the case of an imminent threat, to act without prior consultation to protect agency systems. S. 2151 would provide DHS a much more limited role, requiring it to conduct ongoing security analyses using information provided by the agencies. S. 3342 would give that responsibility instead to OMB.
- Only H.R. 4257 would retain the current FISMA provision giving OMB responsibility for ensuring operation of a federal incident center. However, S. 2105 and the *White House Proposal* each contain other provisions that would establish centers within DHS that would provide for incident reporting, information sharing, and other cybersecurity activities. S. 2151 and S. 3342, in contrast, contain provisions to facilitate reporting to a number of centers (see "Information Sharing" below).

Protection of Privately Held Critical Infrastructure (CI)

The federal government has identified 18 sectors of critical infrastructure (CI),[32] much of which is owned by the private sector. The federal role in protection of privately held CI has been one of the most contentious issues in the debate about cybersecurity legislation. There appears to be broad agreement that additional actions are needed to address the cybersecurity risks to CI,[33] but there is considerable disagreement about how much, if any, additional federal regulation is required. Four of the proposals in *Table 1* address protection of privately held CI.

Both S. 2105 and the *White House Proposal* would require the Secretary of Homeland Security to

- designate as covered CI those private-sector CI entities for which a successful cyberattack could have debilitating or catastrophic impacts of national significance,[34]
- determine what cybersecurity requirements or frameworks are necessary to protect them,
- determine whether additional regulations are necessary to ensure that the requirements are met,
- develop such regulations in consultation with government and private-sector entities, and
- enforce the regulations.

The regulations proposed by S. 2105 would require CI owners and operators, unless exempted,[35] to certify compliance annually, based on self- or third-party assessments, and would provide civil penalties for noncompliance. The Secretary would also be authorized to perform assessments where risks justify such action.

The *White House Proposal* would require owners and operators of covered entities, unless exempted,[36] to submit and attest to compliance plans, and certify compliance annually. Independent evaluations would be performed on a schedule determined by the Secretary. Civil penalties, shutdown orders, and requirements for use of particular measures would be prohibited as enforcement methods.

The *Task Force Report* recommends that Congress consider targeted and limited additional regulation of highly regulated industries where required to improve cybersecurity, and that existing regulations be streamlined. For most CI, however, the report recommends that Congress adopt a menu of voluntary incentives.[37] It also recommends limitations on liability for entities that comply. S. 2105 and the *White House Proposal* would also limit liability for entities in compliance.

The subcommittee version of H.R. 3674[38] would have amended the HSA to require the Secretary of Homeland Security to perform continuous risk assessments of CI for inclusion annually in the National Infrastructure Protection Plan.[39] It would also have required relevant federal regulatory agencies to review cybersecurity regulations for covered CI (as determined by the Secretary[40]) and fill any gaps using a collection of recognized consensus standards, where applicable, and to work with NIST to develop such standards where necessary. It would have prohibited additional regulatory authority beyond the collected standards.

The full-committee version of H.R. 3674[41] would amend the HSA in a substantially different way from the subcommittee version. It would permit the Secretary to engage in risk assessments and other protective activities with respect to privately held CI only upon request by owners and operators. It would require the Secretary to develop a cybersecurity strategy for CI systems and stipulates that the bill would not provide additional authority to DHS over federal or nonfederal entities.

Information Sharing

Barriers to the sharing of information on threats, attacks, vulnerabilities, and other aspects of cybersecurity—both within and across sectors—have long been considered by many to be a significant hindrance to effective protection of information systems, especially those associated with CI.[42] Examples have included legal barriers, concerns about liability and misuse, protection of trade secrets and other proprietary business information, and institutional and cultural factors—for example, the traditional approach to security tends to emphasize secrecy and confidentiality, which would necessarily impede sharing of information.

Proposals to reduce or remove such barriers, including provisions in bills in *Table 1*, have raised concerns,[43] some of which are related to the purpose of barriers that currently impede sharing. Examples include risks to individual privacy and even free speech and other rights, use of information for purposes other than cybersecurity, such as unrelated government regulatory actions, commercial exploitation of personal information, or anticompetitive collusion among businesses that would currently violate federal law (see "Antitrust Laws and Section 5 of the Federal Trade Commission Act").

Five proposals in *Table 1* have provisions for improving information sharing and addressing privacy and other concerns, with H.R. 2674 amending the HSA and H.R. 3523 amending the National Security Act of 1947:

- *Create entities for information sharing.* S. 2105 would require the Secretary of Homeland Security to establish a process for designating federal and nonfederal information exchanges, including a lead federal exchange responsible for facilitating information sharing among federal and nonfederal entities. The *Task Force Report* recommends establishment of a nongovernmental clearinghouse for sharing cybersecurity information among private-sector and government entities. The subcommittee version of H.R. 3674 would have created such an organization, the National Information Sharing Organization (NISO).[44] However, those provisions were omitted from the committee version, which would instead provide statutory authorization for and specify governance and responsibilities of the DHS National Cybersecurity and Communications Integration Center (NCCIC),[45] which was established administratively in 2009.[46] S. 2151 and S. 3342 would not authorize any new entities but list a set of existing centers to which their information-sharing provisions would apply. The DHS center that the *White House Proposal* would establish (see "DHS Authorities for Protection of Federal Systems") would have information sharing as one of its responsibilities.
- *Establish provisions for sharing classified information.* The *Task Force Report*, H.R. 3523, S. 2105, S. 2151, and S. 3342 would establish procedures to permit sharing of classified cybersecurity information with private-sector entities that meet specific criteria.
- *Establish authority for information sharing by and with private-sector entities.*
  - H.R. 3523 would permit cybersecurity providers or self-protected entities to share threat information with other designated entities, notwithstanding any other provision of law. Federal agencies receiving such information would be required to share it with NCCIC, which could share it with other federal entities upon request of the provider of the information.
  - S. 2105 would expressly permit disclosure of lawfully obtained threat indicators among private-sector entities, with the exchanges the bill would establish, and by federal entities with other relevant

federal or private entities, notwithstanding any other provision of law.

- S. 2151 and S. 3342 would permit nonfederal entities to share threat information with cybersecurity centers or with other nonfederal entities for the purpose of addressing threats. S. 2151 would require providers of communications, remote computing, and cybersecurity services under federal contracts to share with cybersecurity centers, through the contracting agency, any threat information related to the contract. S. 3342 would instead require a coordinated process through which providers would inform federal entities of significant incidents with impacts on their missions, with the entity reporting the information to a cybersecurity center. S. 2151 would permit centers to disclose threat information for specified purposes to federal entities, service providers, and nonfederal government entities, whereas S. 3342 would not permit centers to disclose such information to service providers.
- The *White House Proposal* would permit nonfederal entities to disclose information to a designated cybersecurity center for purposes of protection from cybersecurity threats and would permit federal agencies to disclose such information to relevant private entities.

- *Limit disclosure of shared information.* The *Task Force Report*, the subcommittee version of H.R. 3674, H.R. 3523, S. 2105, S. 2151, S. 3342, and the *White House Proposal* would all provide exemptions from the "Freedom of Information Act (FOIA)" for cybersecurity information.[47] All would also restrict disclosure in other ways, such as expressly requiring that it be for specified cybersecurity purposes, although specific requirements vary.
- *Limit government use of information to specified purposes.* The *Task Force Report*, H.R. 3523, H.R. 3674, S. 2151, and S. 3342 would expressly restrict or prohibit regulatory use of shared information. S. 2105 and the *White House Proposal* would limit use of acquired information to cybersecurity or law enforcement purposes. In addition to those uses, H.R. 3523, S. 2151, and S. 3342 would permit use for national security, and H.R. 3523 would add protection from physical harm and, for minors, from pornography or other sexual exploitation.
- *Limit liability for information sharing.* The *Task Force Report*, H.R. 3523, S. 2105, S. 2151, S. 3342, and the *White House Proposal* would protect nonfederal entities from liability for information shared or

other specified actions taken in accordance with the provisions in the legislative proposal. H.R. 3523 would also provide for limited liability for federal violations of restrictions in the bill on disclosure, use, and protection of shared information. The subcommittee version of H.R. 3674 would have permitted actual and punitive civil damages against persons who disclose or use for purposes other than cybersecurity the information that is disclosed to private entities.

- *Provide privacy and civil liberties protections.* All five proposals call for privacy protections. The *Task Force Report* recommends that in providing safe harbors for entities involved in information sharing, "the protection of personal privacy should be at the forefront" (p. 7). It also recommends that the proposed nongovernmental clearinghouse have a privacy board.
  - H.R. 3523 would permit the federal government to "undertake reasonable efforts to limit the impact on privacy and civil liberties" of shared information and require the Inspector General of the Intelligence Community to include, in an annual report to Congress, metrics on impacts of sharing on privacy and civil liberties.[48] It would also require "appropriate" anonymization of shared information.[49] In addition, the bill would prohibit federal use of identifying information from specified sets of library, sales, tax, education, or medical records.
  - The subcommittee version of H.R. 3674 would require that two members of the NISO board of directors be representatives from the privacy and civil liberties community (the committee version), that the NISO charter and procedures include privacy and civil liberties protections, and that anonymization procedures, such as removal of personally identifiable information, be used for shared information. The committee version would create a similar board for the NCCIC and would require ongoing review by the DHS privacy officer of departmental policies and activities.
  - S. 2105 would require the director of the DHS center to appoint a privacy officer, create guidelines for protection of privacy and civil liberties, and ensure that center activities comply with federal requirements. The bill would also require the Secretary of Homeland Security to develop policies and procedures to minimize the impacts of information sharing involving the exchanges that would be established by the bill. It would require three relevant reports: (1) an annual joint report to Congress by the

DHS and Department of Justice privacy officers assessing impacts, (2) a report from the Privacy and Civil Liberties Oversight Board[50] assessing impacts and recommending statutory changes; and (3) a joint report by the Secretary of Homeland Security, the Director of National Intelligence, the Attorney General, and the Secretary of Defense that would include disclosure of significant noncompliance by nonfederal entities with the requirements of the information sharing title of the bill, especially with respect to privacy and civil liberties, with recommendations for any statutory changes.

- S. 2151 would require the heads of agencies with cybersecurity centers to jointly develop procedures for sharing information. Those would consider the need for protection of privacy and civil liberties through anonymization and other means. S. 3342 would in addition permit efforts to limit impacts from sharing on privacy and civil liberties. Both bills would also require biennial joint implementation reports from the agency heads, including review of how shared information may impact privacy and civil liberties, the adequacy of steps to reduce such impact, and any recommended changes to authorities.
- The White House Proposal would require that "reasonable efforts" be taken "to remove information that can be used to identify specific persons unrelated to the cybersecurity threat."51 It would add a new Sec. 248 to the HSA on privacy and civil liberties relating to cybersecurity. It would require the Secretary of Homeland Security, in consultation with privacy and civil liberties experts, to develop and periodically review policies and procedures on information access, disclosure, and use. The policies and procedures would be required to minimize impacts on privacy and civil liberties, safeguard identities, protect confidentiality as much as possible, and provide limits on access, use, and disclosure of information. Agency heads would be required to develop policies for handling information associated with specific persons, to establish programs to monitor and oversee compliance with DHS and agency policies, and to develop and enforce sanctions for violations by agency personnel. The above policies and procedures would be subject to review and approval by the Attorney General. Like S. 2105, the White House Proposal would require an annual joint report to Congress by the

DHS and Department of Justice privacy officers assessing impacts, and a report from the Privacy and Civil Liberties Oversight Board assessing impacts and recommending statutory changes.

Other Topics

Cybercrime Law. S. 2151, S. 3342, the *White House Proposal,* and the *Task Force Report* would each revise current criminal statutes relating to cybersecurity, including criminalizing the damaging of computers associated with critical infrastructure (CI).[52]

Data Breach Notification. The *White House Proposal* and the *Task Force Report* would also both set federal requirements for data breach notification—public notification in cases where a security breach poses significant risks of exposure of sensitive personal information.

Some proposals address additional topics not discussed in this overview. For example, H.R. 2096 would require NIST to develop a strategy for federal use of cloud computing. The *White House Proposal* would restrict the power of state and local governments to require business entities to locate data centers within the state or locality. To the extent that such topics are addressed by amending current statutes, they are discussed below under the relevant laws.

DISCUSSION OF PROPOSED REVISIONS OF CURRENT STATUTES

To identify laws that might be considered candidates for revision, CRS conducted a broad search, consulting with various experts and examining various sources, including legislative proposals in the 111th and 112th Congresses. That search yielded more than 50 potentially relevant statutes (see *Table 2*), of which proposed revisions were identified for 31.[53] For each of the latter group, the report contains an entry that includes

- the popular name of the statute;[54]
- the public law number, along with Statutes-at-Large and relevant U.S. Code citations;[55]
- a brief description of the relevance of the statute for cybersecurity;[56] and

- discussion of potential revisions or updates that have been suggested.[57]

Entries are in chronological order.[58] The statutes discussed include only those for which CRS identified specific proposals to revise them from various observers and in public sources.[59] It does not include proposals for new provisions of federal law that were not identified explicitly as revisions of current named statutes.

One example is the recommendations for statutory language on data-breach notification in the *White House Proposal* and the *Task Force Report.* Neither those two documents, nor the bills on the issue that have been introduced in the 112th Congress,[60] specify named statutes to be revised. One of those bills, S. 1151, would revise 18 U.S.C. Chapter 47 (Fraud and False Statements) by adding a new section at the end, but that provision does not modify any named statute specified either in the bill or in the U.S. Code. It is therefore not included in the discussion below.

However, the bill would also revise 18 U.S.C. §1030, which was added by the "Counterfeit Access Device and Computer Fraud and Abuse Act of 1984," so that provision is discussed.

Another example is bills with provisions clearly related to a named statute, but that do not explicitly modify that statute.

One example from the 111th Congress is H.R. 5590, which had cybersecurity provisions that might be interpreted as modifications to the HSA but were not cited as such. Such provisions are not discussed in this report because their effects on specific statutes could not be determined with certainty.

The approach taken in this report of focusing on statutes by their popular names is useful in many cases, but it has some significant limitations, particularly with respect to the U.S. Code. Some laws, such as the USA Patriot Act of 2001 (see *Table 2*), may be classified across many titles and sections,[61] which may make analysis more challenging. Fortunately, that did not prove to be a significant concern for this report.

However, lack of correspondence between named laws and proposed modification of provisions in the U.S. Code, described above, may in some cases result in significant gaps in coverage of relevant provisions of law relating to cybersecurity by an approach such as the one taken here.

Therefore, the analysis presented here should not be regarded as complete.

Posse Comitatus Act of 1879

Ch. 263, 20 Stat. 152. 18 U.S.C. §1385.[62]

Major Relevant Provisions

- Restricts the use of military forces in civilian law enforcement within the United States, unless it is within a federal government facility.[63]
- Courts have ruled that violations of the act occur when civilian law enforcement makes "direct active use" of military investigators, when use of the military pervades the activities of the civilian officials, or when the military is used so as to subject citizens to military power that is regulatory, prescriptive, or compulsory in nature.

Possible Updates

- Some observers claim that the act prevents the military from cooperating on cybersecurity with civil agencies that may lack the resident expertise and capabilities of the military and DOD.[64] In addition, it may sometimes be difficult to distinguish a criminal cyber attack from one involving national defense, especially if the attack is on a component of critical infrastructure.
- Some have therefore proposed that the act be amended to clarify when U.S. military can operate domestically regarding cyber threats to such infrastructure, most of which is privately owned. Others maintain that no revision is needed because the President has the authority under current law to direct the military to support civil authorities in the event of a domestic disaster.
- A memorandum of agreement signed between DHS and DOD may increase the likelihood that the military would play a significant role in responding to a major cyber attack on U.S. information networks.[65] However, some argue that the defense of U.S. information systems should be solely the purview of civilian agencies such as DHS and the FBI, because involvement of the military creates unacceptable privacy and civil liberties concerns.

Antitrust Laws and Section 5 of the Federal Trade Commission Act

Sherman Antitrust Act

Ch. 647, 26 Stat. 209. 15 U.S.C. §§1-7.

Wilson Tariff Act

Ch. 349, §73, 28 Stat. 570. 15 U.S.C. §§8-11.

Clayton Act

P.L. 63-212, 38 Stat. 730. 15 U.S.C. §§12-27.

Section 5 of the Federal Trade Commission Act (FTC Act)

Ch. 311, §5, 38 Stat. 719. 15 U.S.C. §45(a).[66]

When referred to in statute, the term "antitrust laws" generally means the three laws listed in 15 U.S.C. §12(a), which are the first three statutes listed above. Also frequently included in the list of antitrust laws is Section 5 of the FTC Act, which prohibits unfair and deceptive trade practices. Section 5 is included because courts have found that unfair competition includes, at the least, activity that would violate the Sherman or Clayton Acts.[67]

Major Relevant Provisions

- The antitrust laws as well as Section 5 of the FTC Act are a collection of statutes that forbid combinations or agreements that unreasonably restrain trade.[68] Whenever competitors in a given market share information, antitrust concerns may be raised due to the risk of collusion among competitors.[69]

Possible Updates

Information sharing agreements between private corporations may be subject to antitrust scrutiny, because the sharing of information among competitors could create opportunities for collaboration with the goal of restraining trade.[70] However, information sharing agreements to combat cybersecurity may be in compliance with antitrust principles so long as their goals are to combat cyber threats rather than restrain competition.[71]

Some may argue that in order to develop effective and efficient information sharing agreements to combat cybersecurity threats, an explicit exemption from the antitrust laws for these agreements is necessary. Congress has previously proposed such an exemption. For example, H.R. 2435 (107th Congress) would have granted an express exemption from the antitrust laws and from Section 5 of the FTC Act to persons making and implementing agreements entered into solely for the purpose of "facilitating the correction or avoidance of a cyber security-related problem or communication of or disclosing information to help correct or avoid the effects of a cyber security-related problem." Such an exemption, if enacted by Congress, would allow market participants to engage in information sharing for the purposes of combating cybersecurity threats without concern for implicating the antitrust laws. In the 112th Congress, the *Task Force Report* states that an antitrust exemption *might* be required.[72] H.R. 3523 does not specifically mention antitrust laws, but it permits sharing of cybersecurity information among private-sector entities "notwithstanding any other provision of law." S. 2151 and S. 3342 would expressly exempt from antitrust laws the exchange among private entities of information relating to cybersecurity threats.

Others may argue that the antitrust laws are flexible in nature, particularly as they relate to information sharing agreements, and the laws are flexibly applied by the agencies of jurisdiction.[73] This flexible nature may obviate the need for express exemptions from the application of the laws, while keeping the antitrust agencies involved in and aware of the information sharing agreements companies are making.[74] The agencies have expressed a view that if competitors are collaborating for reasons that do not restrain trade or hamper competition, and safeguards are in place to prevent such restraint, the antitrust laws should not hinder such collaboration.[75] The Department of Justice (DOJ) currently allows companies wishing to create information sharing arrangements for permissible and procompetitive purposes to submit their plans for collaboration to the agency.[76] The agency then reviews the plans and, if the plans are approved, issues what is known as a business review letter.[77] The business review letter will generally state that DOJ does not intend to enforce the antitrust laws against the proposed collaboration. DOJ has issued business review letters to companies who have developed plans to share information to combat cybersecurity threats.[78]

National Institute of Standards and Technology Act

Ch. 872, 31 Stat. 1449.
15 U.S.C. §271 et seq.

Major Relevant Provisions

The original act gave the agency responsibilities relating to technical standards. Later amendments added more generally relevant provisions and, more specifically,

- Identified relevant research topics, among them computer and telecommunication systems, including information security and control systems.[79]
- Established a computer standards program at the National Institute of Standards and Technology (NIST).[80]

Possible Updates

Despite NIST's current authority to conduct research on computers and information security, some concerns have been raised about whether those activities should be enhanced in light of the evolving threat environment for cybersecurity. In the 111th Congress, H.R. 4061, which was passed by the House, would have required NIST to conduct intramural research on identity management and the security of information systems, networks, and industrial control systems. A similar bill, H.R. 2096, is being considered by the 112th Congress.

Federal Power Act

Ch. 285, 41 Stat. 1063.
16 U.S.C. §791a et seq., §824 et seq.[81]

Major Relevant Provisions

- Established the Federal Energy Regulatory Commission (FERC) and gave it regulatory authority over interstate sale and transmission of electric power.

Possible Updates

Concerns about the vulnerability of the electric grid to cyber attack have increased substantially over the last several years.82 Although the Energy Policy Act of 2005 (P.L. 109-58) gave FERC responsibility for developing reliability standards for power systems, limitations to that authority and to the usefulness of the standards-development process to respond effectively to rapidly emerging cybersecurity threats have raised concerns about the need for enhancing FERC's authority to address those threats, especially in light of the development of smart-grid technology.83 Several bills were introduced in the 111th Congress (H.R. 2165, H.R. 2195, H.R. 5026, S. 946, S. 1462) in response. H.R. 5026, which was passed by the House, would have expanded FERC's jurisdiction over electric infrastructure and authorized FERC to order actions by relevant entities in response to threats to cybersecurity. In the 112th Congress, S. 1342 would also provide expanded cybersecurity authorities to FERC, and H.R. 668 would give FERC emergency authorities in response to events causing large-scale disruptions of the electric grid.

Communications Act of 1934

Ch. 652, 48 Stat. 1064. 47 U.S.C. §151 et seq.[84]

Major Relevant Provisions

- Established the Federal Communications Commission (FCC) and gave it regulatory authority over both domestic and international commercial wired and wireless communications.
- Provides the President with authority in a national emergency to control "any or all stations or devices capable of emitting electromagnetic radiations," and in case of war or threat of war, to close "any facility or station for wire communication" (Sec. 706 of the act, 47 U.S.C. §606).

Possible Updates

Some observers have proposed that the act should be revised to give the FCC more of a role in cybersecurity, especially given the growing merging of information and communications technology (ICT) and their increasing importance in the U.S. economy. In fact, a number of other countries have more unified governance of ICT than the United States.[85]

Some controversy exists about whether the Section 706 authorities described above permit the President to shut down Internet communications during a war or national emergency, a power that has sometimes been referred to as the "Internet kill switch."[86] However, there does not appear to be a consensus about whether in fact such additional authority is needed, or, if it is not, whether additional legislation is needed to clarify and delimit it.

That debate became acute during Senate consideration of S. 773 and S. 3480 in the 111th Congress. Those bills would have authorized emergency measures by the President if the operation of critical infrastructure were threatened by cyber attack. A similar provision has been proposed in S. 413 in the 112th Congress.87 This bill also contains a provision that would expressly deny the federal government of any authority to "shut down the Internet."

National Security Act of 1947

Ch. 343, 61 Stat. 495 50 U.S.C. 401 et seq.

Major Relevant Provisions

- Provided the basis for the modern organization of U.S. defense and national security by reorganizing military and intelligence functions in the federal government.
- Created the National Security Council, the Central Intelligence Agency, and the position of Secretary of Defense.
- Established procedures for access to classified information.

Possible Updates

A broad consensus exists that a significant barrier to improving cybersecurity is limitations on sharing of information, including classified information, about cyber-threats and attacks.[88] H.R. 3523 would address that concern by amending the act to facilitate sharing of intelligence information relating to cybersecurity, including classified information, between federal intelligence entities and private-sector providers of cybersecurity services, and to facilitate the identification and sharing of threat information by providers. The bill also includes provisions for protection from liability for entities sharing information and exemption from disclosure of that information under the "Freedom of Information Act (FOIA)."

See also "Information Sharing."

U.S. Information and Educational Exchange Act of 1948 (Smith-Mundt Act)

Ch. 36, 62 Stat. 6.
22 U.S.C. §1431 et seq. [89]

Major Relevant Provisions

- Restricts the State Department from disseminating public diplomacy information domestically and limits its authority to communicate with the American public in general (22 U.S.C. §1461-1a).[90] The domestic dissemination provision originally applied to the now defunct U.S. Information Agency (USIA), which was abolished and its functions transferred to the Secretary of State by P.L. 105-277 (22 U.S.C. §6532).[91]

Possible Updates

Critics maintain that the law is a Cold War relic intended only to restrict the USIA, which no longer exists, from propagandizing Americans with public diplomacy and information materials that were intended for a foreign audience. Those critics argue that the restrictions were created before the advent of the Internet, and the provisions create an obsolete barrier that serves only to prevent the State Department from communicating effectively. Some have also argued that the law has been interpreted to prohibit the military from conducting information operations in cyberspace, as some of those activities could be considered propaganda that could reach U.S. citizens, since the United States does not restrict Internet access according to territorial boundaries.

Yearly appropriations bills for both the State Department and Department of Defense include restrictions on use of funds for "propaganda" activities, although the word "propaganda" is not defined. In the 111th Congress, H.R. 5729 would have removed the so-called "firewall" between domestic and foreign audiences by explicitly authorizing the Department of State to disseminate information through the Internet and information media, stating that the resolution shall "not be construed to prohibit the Department from engaging in any medium of information on a presumption that a U.S. domestic audience may be exposed to program material." However, this provision would have applied only to the State Department; it would not have included DOD or other federal departments or agencies.

State Department Basic Authorities Act of 1956

Ch. 841, 70 Stat. 890. 22 U.S.C. §2651a.

Major Relevant Provisions

- Specifies the organization of the Department of State, including the positions of coordinator for counterterrorism and for HIV/AIDS response.

Possible Updates

As the Internet becomes increasingly international, concerns have been raised about the development and coordination of international efforts in cybersecurity by the United States.[92] In the 111th Congress, S. 3193 would have addressed those concerns by establishing a coordinator for cyberspace and cybersecurity issues within the Department of State. S. 1426 in the 112th Congress contains a similar provision.

Freedom of Information Act (FOIA)

P.L. 89-487, 80 Stat. 250. 5 U.S.C. §552. [93]

Major Relevant Provisions

- Enables any person to access—without explanation or justification—existing, identifiable, unpublished executive-branch agency records, unless the material falls within any of FOIA's nine categories of exemption from disclosure.

Possible Updates

Sharing of cybersecurity information between the federal government and nonfederal entities is widely considered to be an essential need, especially with respect to the protection of critical infrastructure (CI). However, attempts to encourage the private sector to share sensitive CI information with the federal government have, at times, been met with concerns that such records could be subject to public release under FOIA, resulting in potential economic or other harm to the source.

Among the nine exemptions that permit agencies to withhold applicable records are three that may particularly apply to cybersecurity information:

- *Exemption 1:* information properly classified for national defense or foreign policy purposes as secret under criteria established by an executive order.
- *Exemption 3:* data specifically exempted from disclosure by a statute other than FOIA if that statute meets criteria laid out in FOIA.[94]
- *Exemption 4:* trade secrets and commercial or financial information obtained from a person that is privileged or confidential.[95]

An example of Exemption 3 is Sec. 214 of the HSA (see p. 40), which exempts information about the security of critical infrastructure and protected systems that is voluntarily submitted to an agency covered under the act, provided that the entity that supplies the information expressly requests the exemption concurrently.

Despite these existing protections, some private-sector entities may still have concerns about public release of sensitive records—that existing laws may not be specific enough to protect particular types of records, or they may be too narrow to protect all records of concern. The *White House Proposal* would address such concerns by applying Exemption 3 to any lawfully obtained information provided to DHS for cybersecurity purposes.[96] The *Task Force Report* also suggests that a FOIA exemption may be needed,[97] and several bills, including H.R. 3523, S. 2105, S. 2151, and S. 3342 http://www.congress.gov/cgi-lis/bdquery/z?d112:S.2151:, would provide such a FOIA exemption, although none of those proposals would directly modify the statute. Adding such broad exemptions to FOIA, however, could nevertheless prompt concerns about decreases in federal transparency.

Omnibus Crime Control and Safe Streets Act of 1968

P.L. 90-351, 82 Stat. 197.
42 USC Chapter 46, §§3701 to 3797ee-1.

Major Relevant Provisions

- Title I established federal grant programs and other forms of assistance to state and local law enforcement.
- Title III is a comprehensive wiretapping and electronic eavesdropping statute that not only outlawed both activities in general terms but that

also permitted federal and state law enforcement officers to use them under strict limitations.[98]

Possible Updates

The incidence of cybercrime has increased dramatically over the last decade.[99] State and local law enforcement agencies play an important role in combating cybercrime, but concerns have been raised about their abilities to invest sufficient resources in enforcement activities. In the 111th Congress, H.R. 1292 would have added a program for law enforcement grants to state and local criminal justice agencies and relevant nonprofit organizations to combat "white collar crime," including cybercrime.

Racketeer Influenced and Corrupt Organizations Act (RICO)

P.L. 91-452, 84 Stat. 941.
18 U.S.C. Chapter 96, §§1961-1968.

Major Relevant Provisions

- Enlarges the civil and criminal consequences of a list of state and federal crimes when committed in a way characteristic of the conduct of organized crime (racketeering).[100]

Possible Updates

The *Task Force Report* recommends that Congress change RICO "to include computer fraud within the definition of racketeering." [101] The *White House Proposal* would make felony violation of 18 U.S.C. §1030 (see "Counterfeit Access Device and Computer Fraud and Abuse Act of 1984") a racketeering predicate offense.

Federal Advisory Committee Act (FACA)

P.L. 93-579, 86 Stat 770. 5 U.S.C. App., §§1-16.

Major Relevant Provisions

- Specifies the circumstances under which a federal advisory committee can be established, and its responsibilities and limitations.

- Requires that meetings of such committees be open to the public and that records be available for public inspection.[102]

Possible Updates

The act has been criticized as potentially impeding the full development of public/private partnerships in cybersecurity, particularly with respect to impeding private-sector communications and input on policy.[103] While Sec. 871 of the HSA provides the Secretary of Homeland Security with the power to establish advisory committees that are exempt from the requirements of the act, it is possible that additional exemption authority would be helpful. Any such potential benefits might, however, need to be weighed against the impact of such authority on the public's ability to participate in and access the records of affected advisory committees. The draft bill being considered by the House Committee on Homeland Security[104] would exempt the organization created by the bill from requirements of the act.

Privacy Act of 1974

P.L. 93-579, 88 Stat. 1896. 5 U.S.C. §552a.

Major Relevant Provisions

- Limits the disclosure of personally identifiable information (PII) held by federal agencies.
- Requires agencies to provide access to persons with agency records containing information on them.
- Established a code of fair information practices for collection, management, and dissemination of records by agencies, including requirements for security and confidentiality of records.

Possible Updates

Some observers argue that the act should be revised to clarify, in the context of cybersecurity, what is considered PII and how it can be used, such as by explicitly permitting the sharing among federal agencies—or with appropriate third parties such as owners and operators of critical infrastructure—of certain information, such as a computer's Internet (IP) address, in examinations of threats, vulnerabilities, and attacks. The act contains some exemptions, such as for law enforcement activities (5 U.S.C. §552a(b)(7)) and duties of the Comptroller General (5 U.S.C. §552a(b)(10)), but none relating specifically to cybersecurity. However, other observers may argue that the provisions in the act are sufficient to permit necessary cybersecurity activities, and that revising the act to provide additional authorities relating to cybersecurity could compromise the protections provided by the act.[105] In the 112th Congress, H.R. 1732 would revise the act to take changes in information technology into account, but does not specifically address information relating to cybersecurity.

Counterfeit Access Device and Computer Fraud and Abuse Act of 1984

P.L. 98-473, 98 Stat. 2190. 18 U.S.C. §1030.

Major Relevant Provisions

As amended,[106]

- Provides criminal penalties, including asset forfeiture, for unauthorized access and wrongful use of computers and networks of the federal government or financial institutions, or in interstate or foreign commerce or communication;
- Specifies wrongful use as obtaining protected information, damaging or threatening to damage a computer, using the computer to commit fraud, trafficking in stolen computer passwords, and espionage;
- Criminalized electronic trespassing on and exceeding authorized access to federal government computers; and
- Created a statutory exemption for intelligence and law enforcement activities.[107]

Possible Update

The *White House Proposal* would add penalties for damaging certain critical infrastructure computers, increase penalties for most violations of the act, clarify certain offenses, and modify the act's conspiracy and forfeiture provisions. In the 112th Congress, S. 2111, S. 2151, and S. 3342 have similar provisions. S. 890, S. 2151, S. 3342, and the White House Proposal would enlarge the scope of the password trafficking offense by removing the requirement that the computer affect interstate commerce or be used by the United States. S. 1151 would also make several changes similar to but not as extensive as those in the Administration proposal.[108] The *Task Force Report* recommends that the act be broadened to cover critical infrastructure systems, and possibly all private-sector computers, with increased criminal penalties. It also recommends that provisions should be focused narrowly enough to avoid creating unintended liability for legitimate activities.[109]

Electronic Communications Privacy Act of 1986 (ECPA)

P.L. 99-508, 100 Stat. 1848.

18 U.S.C. §§2510-2522, 18 U.S.C. §§2701-2712, 18 U.S.C. §§3121-3126.[110]

Major Relevant Provisions

- Attempts to strike a balance between the fundamental privacy rights of citizens and the legitimate needs of law enforcement with respect to data shared or stored in various types of electronic and telecommunications services.[111] Since the act was passed the Internet and associated technologies have expanded exponentially.[112] The act consists of three parts:
  - A revised Title III of the "Omnibus Crime Control and Safe Streets Act of 1968" (also known as "Title III" or the "Wiretap Act")113 prohibits the interception of wire, oral, or electronic communications unless an exception to the general rule applies. Unless otherwise provided, prohibits wiretapping and electronic eavesdropping; possession of wiretapping or electronic eavesdropping equipment; use or disclosure of information obtained through illegal wiretapping or electronic eavesdropping; and disclosure of information secured through court-ordered wiretapping or electronic eavesdropping, in order to obstruct justice.[114]
  - The Stored Communications Act (SCA)115 prohibits unlawful access to stored communications.[116]
  - The Pen Register and Trap and Trace statute governing the installation and use of trap and trace devices and pen registers,117 proscribing unlawful use of a pen register or a trap and trace device.[118]

- Establishes rules that law enforcement must follow before they can access data stored by service providers. Depending on the type of customer information involved and the type of service being provided, the authorization law enforcement must obtain in order to require disclosure by a third party will range from a simple subpoena to a search warrant based on probable cause.

Possible Updates

ECPA reform efforts focus on crafting a legal structure that is up-to-date, can be effectively applied to modern technology, and that protects users' reasonable expectations of privacy. ECPA is viewed by many stakeholders as unwieldy, complex, and difficult for judges to apply.[119] Cloud computing[120] poses particular challenges to the ECPA framework. For example, when law enforcement officials seek data or files stored in the cloud, such as web-based e-mail applications or online word processing services, the privacy standard that is applied is often lower than the standard that applies when law enforcement officials seek the same data stored on an individual's personal or business hard drive.[121]

An ECPA reform advocacy coalition has advanced the following principles:

- A governmental entity may require an entity covered by ECPA (a provider of wire or electronic communication service or a provider of remote computing service) to disclose communications that are not readily accessible to the public, but only with a search warrant issued based on a showing of probable cause, regardless of the age of the communications, the means or status of their storage or the provider's access to or use of the communications in its normal business operations.
- A governmental entity may access, or may require a covered entity to provide, prospectively or retrospectively, location information regarding a mobile communications device, but only with a warrant issued based on a showing of probable cause.
- A governmental entity may access, or may require a covered entity to provide, prospectively or in real time, dialed number information, e-mail to and from information or other data currently covered by the authority for pen registers and trap and trace devices, but only after judicial review and a court finding that the governmental entity has made a showing at least as strong as the showing under 2703(d).

- Where the Stored Communications Act authorizes a subpoena to acquire information, a governmental entity may use such subpoenas only for information related to a specified account(s) or individual(s). All nonparticularized requests must be subject to judicial approval.[122]

The *Task Force Report* recommends changes to laws governing the protection of electronic communications to facilitate sharing of appropriate cybersecurity information, including the development of an anonymous reporting mechanism.[123]

Department of Defense Appropriations Act, 1987

P.L. 99-591, 100 Stat. 3341-82, 3341-122. 10 U.S.C. §167.[124]

Major Relevant Provisions

Provides specific authority to the U.S. Special Operations Command (USSOCOM) for the conduct of direct action, strategic reconnaissance, unconventional warfare, foreign internal defense, civil affairs, and psychological operations; also counterterrorism, humanitarian assistance, theater search and rescue, and such other activities as may be specified by the President or the Secretary of Defense.

Possible Update

In addition to the authority provided under this act, Title 10 of the U.S. Code provides inherent and specific authority to DOD to undertake the following activities:

- §113 provides that, subject to the direction of the President, the Secretary of Defense has authority, direction, and control over DOD;
- §164 provides specific authority for combatant commanders for the performance of missions assigned by the President or by the Secretary with the approval of the President.

Specific authorities for combatant commanders are provided in Title 10 to use force in self-defense and for mission accomplishment—including in the recently recognized information operations environment. In preparing for contingencies or military operations, DOD undertakes activities to lessen risks to U.S. interests, including discrete actions to prepare for and respond to a cyberwarfare-related incident.[125]

Some military activities are conducted *clandestinely* to conceal the nature of the operation and passively collect intelligence. Activities focused on influencing the governing of a foreign country are deemed *covert* actions[126] and may not be conducted by members of the military absent a presidential finding and notification of the congressional intelligence committees.[127]

Some analysts suggest that in the cyber domain distinguishing between whether an action is or should be considered covert or clandestine is problematic, as an attacking adversary's intent and location are often difficult to discern. Should this act be updated, reassessing DOD's authorities in light of its unique intelligence capabilities may assist in responding to and conducting offensive cyber attacks.

High Performance Computing Act of 1991

P.L. 102-194, 105 Stat. 1594. 15 U.S.C. Chapter 81.[128]

Major Relevant Provisions

- Establishes a federal high-performance computing program and requires that it address security needs.
- Requires that the program provide for interagency coordination and that an annual report on implementation be submitted to Congress.
- Requires NIST to establish security and privacy standards in high-performance computing for federal systems.

Possible Updates

This act established the Networking and Information Technology Research and Development (NITRD) Program, which produces the required annual report. However, concerns have been raised that the program does not yield sufficient strategic planning and does not sufficiently stress cybersecurity research and development (R&D). In the 111th Congress, H.R. 2020, which passed the House, would have addressed that concern by requiring a five-year strategic plan with three-year reviewing cycle. It would also have added a research goal of increasing understanding "of the scientific principles of cyber-physical systems" and improving methods for designing, developing, and operating such systems with high reliability, safety, and security. H.R. 3834 in the 112th Congress is similar but adds provisions on cloud computing. S. 773

in the 111th Congress would have required NIST to develop cybersecurity standards and metrics for computer networks and user interfaces, as would S. 2105 in the 112th Congress. S. 2151 and S. 3342 would establish cybersecurity, including security of supply chains, as one of the goals for research under the act and contains a requirement similar to that of H.R. 3834 for cyber-physical systems.

H.R. 3834, S. 2151, and S. 3342 would also make a number of other amendments not directly related to cybersecurity.

Communications Assistance for Law Enforcement Act of 1994 (CALEA)

P.L. 103-414, 108 Stat. 4279. 47 U.S.C. §1001 et seq.[129]

Major Relevant Provisions

- Requires telecommunications carriers to assist law enforcement in performing electronic surveillance on their digital networks pursuant to court orders or other lawful authorization.
- Directs the telecommunications industry to design, develop, and deploy solutions that meet requirements for carriers to support authorized electronic surveillance, including unobtrusive isolation of communications and call-identifying information for a target and provision of that information to law enforcement, in a manner that does not compromise the privacy and security of other communications.

Possible Updates

Some government and industry observers believe that CALEA should be revised to improve its effectiveness in addressing cybersecurity concerns. Among the concerns expressed are whether the act is the best mechanism for collecting information transmitted via the Internet, whether reassessment is needed of which private-sector entities the act covers and which government entities should be involved in enforcement and oversight, and what the role of industry should be in the development of the technologies and standards used to implement the provisions of the act.

The *Task Force Report* recommends changes to laws governing the protection of electronic communications to facilitate sharing of appropriate cybersecurity information, including the development of an anonymous reporting mechanism.[130]

Communications Decency Act of 1996

P.L. 104-104 (Title V), 110 Stat. 133. 47 U.S.C. §§223, 230.[131]

Major Relevant Provisions

- Intended to regulate indecency and obscenity on telecommunications systems, including the Internet. Although much of the law is targeted at lewd or pornographic material, particularly when shown to children under the age of 18, the obscenity and harassment provisions could also be interpreted as applying to graphic, violent terrorist propaganda or incendiary language.
- Section 230(c)(1) asserts that "no provider or user of an interactive computer service shall be treated as the publisher or speaker of any information." This has been interpreted to absolve Internet service providers and certain web-based services of responsibility for third-party content residing on those networks or websites.[132]

Possible Updates

Some argue that certain Internet content, such as terrorist chat rooms or propaganda websites, presents a national security or operational threat that is not represented within the Communications Decency Act. Further, should such material be deemed as "indecent," the law does not give federal agencies the authority to require that the Internet service providers hosting the content to take it offline.

These critics maintain that the law should be revised to compel ISPs and web administrators to dismantle sites containing information that could be used to incite harm against the United States.

A possible revision could be similar to the "take down and put back" provision in the Digital Millennium Copyright Act, 112 Stat. 2860, P.L. 105-304 which amended title 17 of the U.S. Code to hold a service liable for publishing material that is defamatory or infringes upon a third party copyright.

Others maintain that such a revision is counter to the spirit of free, open exchange of information that is characterized by the Internet and may be a First Amendment violation. Some have also expressed concerns that the intelligence value gained by preserving and monitoring the sites outweighs the potential threat risk.

Clinger-Cohen Act (Information Technology Management Reform Act) of 1996

P.L. 104-106 (Divisions D and E), 110 Stat. 642. 40 U.S.C. §11101 et seq.[133]

Major Relevant Provisions

- Gave agency heads authority to acquire IT and required them to ensure the adequacy of agency information security policies.
- Established the position of agency Chief Information Officer (CIO), responsible for assisting agency heads in IT acquisition and management.
- Requires the Office of Management and Budget (OMB) to oversee major information technology (IT) acquisitions.
- Requires OMB to promulgate, in consultation with the Secretary of Homeland Security, compulsory federal computer standards based on those developed by the National Institute of Standards and Technology (NIST).[134]
- Exempts national security systems from most provisions.

Possible Update

With the increasing globalization of the IT hardware and software industries, concerns have been growing among cybersecurity experts about potential vulnerabilities at various points along the supply chain for IT products. H.R. 1136, introduced in the 112th Congress, would address such concerns with respect to federal acquisition of IT products and services by requiring vendors to meet security requirements to be developed by OMB, and also requiring vulnerability assessments by agencies.

S. 413 (similar to S. 3480 in the 111th Congress), S. 2105, S. 2151, S. 3342, and the *White House Proposal* would return the authority for promulgating standards for federal systems to the Secretary of Commerce.[135] H.R. 4257, in contrast, would not amend that provision.

Congress and the executive branch have debated the limits of the authority and jurisdiction of CIOs since their establishment. In the private sector, CIOs may often serve as the senior IT decision maker. In federal agencies, in contrast, CIOs do not have budgetary control or authority over IT resources.[136] As part of a plan to reform federal IT management,[137] the Obama Administration has indicated its intention to change the role of CIOs "away from just policymaking and infrastructure maintenance, to encompass true portfolio management for all IT," including information security.[138] The *White House Proposal* does not include any provisions related to that proposed change, but additional legislative authority may be required for such a change to be fully implemented.

The Obama Administration also appointed a federal chief information officer and a federal chief technology officer (CTO), positions first created in the George W. Bush Administration, where the OMB deputy director of management also served as federal CIO. In the 111th Congress, H.R. 1910 and H.R. 5136, and H.R. 1136 in the 112th Congress, contain provisions to establish a statutory basis for the CTO position, not, however, explicitly as amendments to the Clinger-Cohen Act.[139] Some proposals in previous Congresses would also have established the federal CIO position in law.[140]

Identity Theft and Assumption Deterrence Act of 1998

P.L. 105-318, 112 Stat. 3007. 18 U.S.C. §1028.[141]

Major Relevant Provisions

- Made identity theft a federal crime.
- Provided penalties for individuals who either committed or attempted to commit identity theft.
- Provided for forfeiture of property used or intended to be used in the fraud.
- Directed the Federal Trade Commission (FTC) to record complaints of identity theft, provide victims with informational materials, and refer complaints to the appropriate consumer reporting and law enforcement agencies.[142]

Possible Updates

See "Identity Theft Penalty Enhancement Act" below.

Homeland Security Act of 2002 (HSA)

P.L. 107-296 (Titles II and III), 116 Stat. 2135. 6 U.S.C. §§121-195c, 441-444, and 481-486.[143]

Major Relevant Provisions

- Transferred some functions relating to the protection of information infrastructure from other agencies to the Department of Homeland Security (DHS).[144]
- Requires DHS to provide state and local governments and private entities with threat and vulnerability information, crisis-management support, and technical assistance relating to recovery plans for critical information systems.
- Permits the Secretary of Homeland Security to designate qualified technologies as subject to certain protections from liability in claims relating to their use in response to an act of terrorism.[145]
- Established mechanisms to facilitate information sharing among federal agencies and appropriate nonfederal government and critical-infrastructure personnel.[146]
- Authorized DHS to establish a system of volunteer experts ("Net Guard") to assist local communities in responding to attacks on information and communications systems.
- Strengthened some criminal penalties relating to cybercrime.
- Created the Directorate of Science and Technology within DHS and assigned it broad R&D responsibilities, although responsibilities relating to cybersecurity R&D were not specifically described.

Possible Updates

Various concerns have been raised about the ways in which the act addressed cybersecurity, and a number of proposals have been made since its enactment to enhance the cybersecurity provisions. In the 111th Congress, the most comprehensive legislative proposal was in S. 3480, which was reported out of the Senate Committee on Homeland Security and Governmental Affairs in the 111th Congress, and reintroduced in the 112th Congress as S. 413 with minor modifications. It would add provisions on cybersecurity that would

- establish a center for cybersecurity and communications within DHS;
- require coordination with the DHS Office of Infrastructure Protection and sector-specific agencies;

- establish the United States Computer Emergency Readiness Team (US-CERT) within the center;
- stipulate information-sharing procedures for federal agencies and other entities;
- establish a program within the center to provide assistance to the private sector;
- require the center to identify cyber vulnerabilities to critical infrastructure and establish requirements to address them;
- establish procedures for response to imminent cyber threats to critical infrastructure,[147] enforcement of requirements, and protection of information; and
- require a risk-management strategy for security of the supply chain.

It would establish a cybersecurity R&D program in DHS and require coordination of those activities with other agencies and private entities. It would also establish a public/private-sector cybersecurity advisory council.

The *White House Proposal* would also substantially enhance DHS authority relating to cybersecurity. The proposal would differ in several ways from the approach taken by S. 413. Among other differences, it would provide enhanced authority to the DHS Secretary that S. 413 provides directly to a new center within the department. However, the *White House Proposal* would require the Secretary to establish a center with cybersecurity responsibilities for federal and critical infrastructure systems.[148] It also does not codify the establishment of US-CERT, unlike S. 413, and does not provide the President with the authority to implement emergency actions in response to an imminent risk to critical infrastructure. It does, however, provide the DHS Secretary with authority to direct responses of federal agencies to cybersecurity threats or incidents.

S. 2105 contains elements of both the *White House Proposal* and S. 413. It would establish a new center, with new authorities, but omits the provision in S. 413 establishing US-CERT by law, as well as the provision on presidential emergency powers. S. 2105 would require the Science and Technology Directorate of DHS to establish a cybersecurity R&D program. S. 1546 would also require departmental cybersecurity research.

H.R. 3674, as reported to the House, would provide additional responsibilities and authorities to DHS for the protection of federal information systems. It would provide for information sharing with federal and nonfederal entities, cybersecurity research and development (R&D), and recruitment and retention of cybersecurity personnel. To facilitate information sharing and technical assistance, it would create a center within DHS that would include a private-sector board of advisors. Unlike the bill as introduced, it does not include a nongovernmental clearinghouse for sharing cybersecurity information between the private sector and the federal government that was recommended by the *Task Force Report*. H.R. 3674 would also require DHS to perform cybersecurity R&D, to include testing, evaluation, and technology transfer.

Some other bills in the 111th Congress would also have revised the act. H.R. 6423, reintroduced as H.R. 174 in the 112th Congress, would establish a new office to develop, oversee, and enforce cybersecurity compliance for critical infrastructure sectors. H.R. 266, reintroduced as H.R. 76, would add a cybersecurity fellowship program for nonfederal officials to familiarize them with DHS cybersecurity activities. H.R. 4507 and H.R. 4842 would have added a cybersecurity training initiative for first responders and others. H.R. 2868 and S. 3599 would have added chemical-facility security measures, including cybersecurity, to the act.

See also "DHS Authorities for Protection of Federal Systems," "Cybersecurity Workforce," "Research and Development," "Protection of Privately Held Critical Infrastructure (CI)," and "Information Sharing."

Federal Information Security Management Act of 2002 (FISMA)

P.L. 107-296 (Title X), 116 Stat. 2259. P.L. 107-347 (Title III), 116 Stat. 2946.

44 U.S.C. Chapter 35, Subchapters II and III, [40 U.S.C. 11331, 15 U.S.C. 278g-3 & 4].[149]

Major Relevant Provisions

FISMA created a security framework for federal information systems, with an emphasis on risk management, and gave specific responsibilities to the Office of Management and Budget (OMB), the National Institute of Standards and Technology (NIST), and the heads, chief information officers (CIOs), chief information security officers (CISOs), and inspector generals (IGs) of federal agencies.[150]

- Required executive agencies to inventory major computer systems, identify and provide appropriate security protections, and develop, document, and implement agency-wide information security programs.

- Gave OMB responsibility for overseeing federal information-security policy and evaluating agency information-security programs, but exempted national security systems, except with respect to enforcement of accountability for meeting requirements and reporting to Congress.
- Revised the responsibilities of the Secretary of Commerce and NIST for information-system standards and transferred responsibility for promulgation of those standards from the Secretary of Commerce to OMB.[151]
- Required that NIST cybersecurity standards be complementary with those developed for national security systems, to the extent feasible.
- Required heads of federal agencies to provide security protections commensurate with risk and to comply with applicable security standards. Specifically required agencies using national security systems to provide security protections commensurate with risk and in compliance with standards for such systems.
- Required senior agency officials to perform risk assessments, to determine and implement necessary security controls in a cost-effective manner, and to evaluate those controls periodically.
- Designated specific information-security responsibilities for agencies' chief information security officers, including agency-wide information-security programs, policies, and procedures, and training of security and other personnel.
- Required designation of an information-security officer in each agency, security awareness training, processes for remedial action to address deficiencies, and procedures for handling security incidents and ensuring continuity of operations.
- Required annual agency reports to Congress, performance plans, and independent evaluations of information security.
- Established a central federal incident center, overseen by OMB, to analyze incidents and provide technical assistance relating to them, to inform agency operators about current and potential threats and vulnerabilities, and to consult with NIST, NSA, and other appropriate agencies about incidents.
- Gave responsibility for protection of mission-crucial systems in DOD and the CIA to the Secretary of Defense and the DCI, respectively,

and required the Secretary of Defense to include compliance with the provisions above in developing program strategies for the Defense Information Assurance Program (10 U.S.C. §2224).

Possible Updates

A commonly expressed concern about FISMA is that it is awkward and inefficient in providing adequate cybersecurity to government IT systems. The causes cited have varied but common themes have included inadequate resources, a focus on procedure and reporting rather than operational security, lack of widely accepted cybersecurity metrics, variations in agency interpretation of the mandates in the act, excessive focus on individual information systems as opposed to the agency's overall information architecture, and insufficient means to enforce compliance both within and across agencies.[152] Several legislative proposals in the 111th and 112th Congresses have included major revisions to the act. The proposals varied in detail, with several notable provisions in some:

- Creation of a White House office with responsibility for cybersecurity;
- Transfer of responsibilities from OMB to the Secretary of Homeland Security or the Secretary of Commerce;
- Revisions to agency responsibilities under the act, including continuous monitoring, use of metrics, and emphasis on risk-based rather than minimum security measures;
- Changes in reporting requirements;
- Specification of cybersecurity requirements for acquisitions and the IT supply chain; and
- Establishment of mechanisms for interagency collaboration on cybersecurity.

In the 111th Congress, H.R. 5136 passed in the House,[153] and S. 3480 was reported out of the Senate Homeland Security and Governmental Affairs Committee.

In the 112th Congress, the *Task Force Report* recommends an increased focus on monitoring, support for DHS authority, and taking new and emerging technologies, such as cloud computing, into account.[154] H.R. 1136 would make many changes similar to those in H.R. 5136 in the 111th Congress, transferring responsibility to a new White House Office for Cyberspace created by the bill. H.R. 4257, in contrast, retains the current role of the OMB Director. H.R. 4257 passed the House under suspension of the rules in April 2012.

S. 413 would make changes similar to those in S. 3480 in the previous Congress, transferring responsibility for federal information security policy from the Director of OMB to the Director of a new DHS center that the bill would establish. The *White House Proposal* is broadly similar to congressional proposals in many details. However, it would not create a White House cybersecurity office and would transfer responsibilities to the DHS Secretary rather than to a new cybersecurity center within DHS. S. 2105 includes a similar approach. S. 2151 and S. 3342, in contrast, would transfer responsibilities from OMB to the Secretary of Commerce.

S. 1535 would require that agency information security programs assess the practices of contractors and third parties with respect to sensitive personally identifiable information as defined in the bill and ensure that any deficiencies are remediated.

See also "FISMA Reform."

Terrorism Risk Insurance Act of 2002

P.L. 107-297, 116 Stat. 2322. 15 U.S.C. §6701 nt.[155]

Major Relevant Provisions

Provides federal cost-sharing subsidies for insured losses resulting from acts of terrorism.

Possible Updates

The act is intended to provide incentives for the development of insurance coverage for losses from acts of terrorism. Losses from cyber attacks are not specifically included, and some observers have raised concerns about whether some modification of the act would be appropriate.[156]

Cyber Security Research and Development Act, 2002

P.L. 107-305, 116 Stat. 2367,
15 U.S.C. [§§278g,h], §7401 et seq.[157]

Major Relevant Provisions

- Requires the National Science Foundation (NSF) to award grants for basic research to enhance computer security and for improving undergraduate and master's degree programs, doctoral research, and faculty development programs in computer and network security; and to establish multidisciplinary centers for research on computer and network security.
- Requires NIST to establish programs to award postdoctoral and senior research fellowships in cybersecurity and to assist institutions of higher learning that partner with for-profit entities to perform cybersecurity research; to perform intramural specified cybersecurity research; and to develop a checklist of security settings for federal computer hardware and software for voluntary use by federal agencies.

Possible Updates

A commonly expressed concern about federal research and development (R&D) relating to cybersecurity has been that it is insufficiently coordinated and prioritized, and focuses too little on understanding of fundamental principles and using them to develop transformational technologies. The George W. Bush Administration attempted to address the latter gap through the "leap-ahead" technology component of the Comprehensive Cybersecurity Initiative.[158] The Obama Administration's policy review[159] also called for expanded, transformational research.

Concerns have also been raised about the need to improve the process by which NIST creates checklists and other guidance and technical standards for federal IT systems.[160]

H.R. 4061 in the 111th Congress would have addressed those concerns by revising the act. A similar bill in the 112th Congress, H.R. 2096, would, as amended, expand NSF R&D programs in cybersecurity, and require NIST to develop automated security specifications for its cybersecurity standards, checklists, and associated data. S. 2105, S. 2151, and S. 3342 would also expand cybersecurity topics addressed by NSF.

E-Government Act of 2002

P.L. 107-347, 116 Stat. 2899.

5 U.S.C. Chapter 37, 44 U.S.C. 3501 nt., 44 U.S.C. Chapter 35, Subchapter 2, and Chapter 36.

Major Relevant Provisions

Serves as the primary legislative vehicle to guide federal IT management and initiatives to make information and services available online. Significant provisions include the following:

- Established the Office of Electronic Government within OMB, to be headed by an administrator with a range of IT management responsibilities, including cybersecurity.
- Established the interagency CIO (Chief Information Officer) Council and specified working with the National Institute of Standards and Technology (NIST) on security standards as one of its functions.
- Assigned agency CIOs responsibility for monitoring implementation of federal cybersecurity standards in their agencies.
- Contains various other requirements for security and protection of confidential information, including electronic authentication and privacy guidelines.
- Established a five-year personnel exchange program between federal agencies and private sector organizations to help agencies fill IT management training needs.
- Also included the "Federal Information Security Management Act of 2002 (FISMA)."

Possible Update

The *White House Proposal* would renew the personnel exchange program, which terminated at the end of 2007, and remove the current restriction in eligibility to management personnel. While this program would be applicable to any subdiscipline of IT, a widely held belief at present is that gaps in cybersecurity expertise are of particular concern. S. 1732 would revise the privacy provisions to account for the increased commercial availability of personally identifiable information, which the bill defines broadly.[161] It would also require agencies to designate chief privacy officers and create a council of them, and broaden OMB's privacy responsibilities.

Identity Theft Penalty Enhancement Act

P.L. 108-275, 118 Stat. 831. 18 U.S.C. §§1028, 1028A.[162]

Major Relevant Provisions

- Established penalties for *aggravated* identity theft, in which a convicted perpetrator could receive additional penalties (two to five years' imprisonment) for identity theft committed in relation to other federal crimes.[163]

Possible Updates

While the number of reported incidents of identity theft fell in 2010, identity theft has generally been the fastest growing type of fraud in the United States over the past decade.[164] FTC complaint data indicate that the most common fraud complaint received (19% of all consumer fraud complaints in 2010) has remained that of identity theft.[165] In 2010, for instance, about 8.1 million Americans were reportedly victims of identity theft. This is a decrease of about 28% from the approximately 11.1 million who were victimized in 2009.[166] Javelin Strategy and Research estimates that identity theft cost consumers about $37 billion in 2010.

The most recent congressional action taken to enhance the identity theft laws was through the Identity Theft Enforcement and Restitution Act of 2008 (Title II of P.L. 110-326). Among other elements, several of which were recommended by a presidential task force in 2007,[167] the act authorized restitution to identity theft victims for their time spent recovering from the harm caused by the actual or intended identity theft. Legislation has not yet, however, adopted recommendations of the task force to

- amend the identity theft and aggravated identity theft statutes so that thieves who misappropriate the identities of corporations and organizations—and not just the identities of individuals—can be prosecuted,[168] and
- amend the aggravated identity theft statute by adding new crimes as predicate offenses[169] for aggravated identity theft violations.[170]

The task force recommended that Congress clarify the identity theft and aggravated identity theft statutes to cover both individuals and organizations targeted by identity thieves because the range of potential victims includes not only individuals but organizations as well. The task force cites "phishing" as a means by which identity thieves assume the identity of a corporation or organization in order to solicit personally identifiable information from individuals.[171]

In part because identity theft is a facilitating crime, and the criminal act of stealing someone's identity often does not end there, investigating and prosecuting identity theft often involves investigating and prosecuting a number of related crimes. In light of this interconnectivity, the task force recommended expanding the list of predicate offenses for aggravated identity theft. The task force specifically suggested adding identity theft-related crimes such as mail theft,[172] counterfeit securities,[173] and tax fraud.[174]

The *Task Force Report* also recommends requiring restitution for victims of identity theft and computer fraud.[175] At present, the statute authorizes restitution but does not require it.

Intelligence Reform and Terrorism Prevention Act of 2004 (IRTPA)

P.L. 108-458, 118 Stat. 3638.

42 U.S.C. §2000ee, 50 U.S.C. §403-1 et seq., §403-3 et seq., §404o et. seq.[176]

Major Relevant Provisions

- Established the position of the Director of National Intelligence.
- Establishes mission responsibilities for some entities in the intelligence, homeland security, and national security communities.
- Discusses issues related to the collection, analysis, and sharing of security-related information.
- Establishes a Privacy and Civil Liberties Board within the Executive Office of the President.

Possible Updates

The act does not contain a single reference to cyber, cybersecurity, or related activities. Its stated purpose is to "reform the intelligence community and the intelligence and intelligence-related activities of the United States Government, and for other purposes." The act contains findings and recommendations offered in the 9/11 Commission Report[177] and other assessments that address national and homeland security shortcomings associated with the terrorist attacks of September 11, 2001.

Numerous organizations, programs, and activities in the act currently address cybersecurityrelated issues. IRPTA addresses many types of risks to the nation and threats emanating from man-made and naturally occurring events. The broad themes of the act could be categorized as how the federal government identifies, assesses, defeats, responds to, and recovers from current and emerging threats. The act might be updated to incorporate cybersecurity-related issues. However, any such update could affect numerous organizations and activities. 178

Table 2. Laws Identified as Having Relevant Cybersecurity Provisions

Year	Popular Name	Law	Stat.	U.S.C.	Applicability and Notes	CRS Reports
6/18/1878	Posse Comitatus Act (p. 20)	Ch. 263	20 Stat. 152	18 U.S.C. §1385	Restricts the use of military forces in civilian law enforcement within the United States. May prevent assistance to civil agencies that lack DOD expertise and capabilities.	RS20590
7/2/1890 and later	Antitrust Laws: (p. 21)					
	Sherman Antitrust Act, Wilson Tariff Act	Ch. 647 Ch. 349,	26 Stat. 209 28 Stat. 570	15 U.S.C. §§1-7 15 U.S.C. §§8-11	"Antitrust laws" generally means the three laws listed in 15 U.S.C. §12(a) and Sec. 5 of the FTC Act, which forbid combinations or agreements that unreasonably restrain trade. May create barriers to sharing of information or collaboration to enhance cybersecurity among private sector entities.	
	Clayton Act	§73				
	Sec. 5 of the Federal	P.L.	38 Stat.	15 U.S.C.		
	Trade Commission (FTC) Act	63-212 Ch, 311, §5	730 38 Stat. 719	§§12-27 15 U.S.C. §45(a)		
3/3/1901	National Institute of Standards and Technology (NIST) Act (p. 23)	Ch. 872	31 Stat. 1449	15 U.S.C. §271 et seq.	The original act gave the agency responsibilities relating to technical standards. Later amendments established a	
					computer standards	

Year	Popular Name	Law	Stat.	U.S.C.	Applicability and Notes	CRS Reports
					program and specified research topics, among them computer and telecommunication systems, including information security and control systems.	
8/13/1912	Radio Act of 1912	Ch. 287	37 Stat. 302		Established a radio licensing regime and regulated private radio communications, creating a precedent for wireless regulation. Repealed by the Radio Act of 1927.	
6/10/1920	Federal Power Act (p. 23)	Ch. 285	41 Stat. 1063	16 U.S.C. §791a et seq., §824 et seq.	Established the Federal Energy Regulatory Commission (FERC) and gave it regulatory authority over interstate sale and transmission of electric power. The move toward a national smart grid is raising concerns about vulnerability to cyber attack.	R41886
2/23/1927	Radio Act of 1927	Ch. 169	44 Stat. 1162		Created the Federal Radio Commission as an independent agency (predecessor of the FCC) and outlawed interception and divulging private radio messages. Repealed by the Communications Act of 1934 (see p. 24).	
6/19/1934	Communications Act of 1934 (p.24)	Ch. 652	48 Stat. 1064	47 U.S.C. §151 et seq.	Established the Federal Communications Commission (FCC) and gave it regulatory authority over both domestic and international commercial wired and wireless communications. Provides the President with emergency powers	RL325 RL346

Table 2. (Continued)

Year	Popular Name	Law	Stat.	U.S.C.	Applicability and Notes	CRS Reports
					over communications stations and devices. Governs protection by cable operators of information about subscribers.	
7/26/1947	National Security Act of 1947 (p. 25)	Ch. 343	61 Stat. 495	50 U.S.C. §401 et seq.	Provided the basis for the modern organization of U.S. defense and national security by reorganizing military and intelligence functions in the federal government. Created the National Security Council, the Central Intelligence Agency, and the position of Secretary of Defense. Established procedures for access to classified information.	
1/27/1948	US Information and Educational Exchange Act of 1948 (Smith-Mundt Act) (p. 25)	Ch. 36	62 Stat. 6	22 U.S.C. §1431 et seq.	Restricts the State Department from disseminating public diplomacy information domestically and limits its authority to communicate with the American public in general. Has been interpreted by some to prohibit the military from conducting cyberspace information operations, some of which could be considered propaganda that could reach U.S. citizens, since the government does not restrict Internet access according to territorial boundaries.	R41674
9/8/1950	Defense Production	Ch. 932	64 Stat. 798	50 U.S.C. App. §2061	Codifies a robust legal authority given the	RS20587 RL31133

Year	Popular Name	Law	Stat.	U.S.C.	Applicability and Notes	CRS Reports
	Act of 1950			et seq.	President to force industry to give priority to national security production and ensure the survival of security-critical domestic production capacities. It is also the statutory underpinning of governmental review of foreign investment in U.S. companies.	
8/1/1956	State Department Basic Authorities Act of 1956 (p. 27)	P.L. 84-885	70 Stat. 890	22 U.S.C. §2651a	Specifies the organization of the Department of State, including the positions of coordinator for counterterrorism. As the Internet becomes increasingly international, concerns have been raised about the development and coordination of international efforts in cybersecurity by the United States.	R40989
10/30/1965	Brooks Automatic Data Processing Act	P.L. 89-306	79 Stat. 1127		Gave GSA authority over acquisition of automatic data processing equipment by federal agencies, and gave NIST responsibilities for developing standards and guidelines relating to automatic data processing and federal computer systems. Repealed by the Clinger-Cohen Act of 1996 (see p. 38).	
7/4/1966	Freedom of Information Act (FOIA) (p. 27)	P.L. 89-487	80 Stat. 250	5 U.S.C. §552	Enables anyone to access agency records except those falling into nine categories of exemption, among them classified documents, those exempted by	R41406 R41933

Table 2. (Continued)

Year	Popular Name	Law	Stat.	U.S.C.	Applicability and Notes	CRS Reports
					specific statutes, and trade secrets or other confidential commercial or financial information.	
6/19/1968	Omnibus Crime Control and Safe Streets Act of 1968 (p. 29)	P.L. 90-351	82 Stat. 197	42 U.S.C. Chapter 46, §§3701 to 3797ee-1	Title I established federal grant programs and other forms of assistance to state and local law enforcement. Title III is a comprehensive wiretapping and electronic eavesdropping statute that not only outlawed both activities in general terms but that also permitted federal and state law enforcement officers to use them under strict limitations.	
10/15/1970	Racketeer Influenced and Corrupt Organizations Act (RICO) (p. 29)	P.L. 91-452	84 Stat. 941	18 U.S.C. Chapter 96, §§1961-1968	Enlarges the civil and criminal consequences of a list of state and federal crimes when committed in a way characteristic of the conduct of organized crime (racketeering).	96-950
10/6/1972	Federal Advisory Committee Act (p. 30)	P.L. 92-463	86 Stat 770	5 U.S.C. App., §§1-16	Specifies conditions for establishing a federal advisory committee and its responsibilities and limitations. Requires open, public meetings and that records be available for public inspection. Has been criticized as potentially impeding the development of public/private partnerships in cybersecurity,	R40520

Year	Popular Name	Law	Stat.	U.S.C.	Applicability and Notes	CRS Reports
					particularly private-sector communications and input on policy.	
11/7/1973	War Powers Resolution	P.L. 93-148	87 Stat. 555	50 U.S.C. Chapter 33, §§1541-1548.	Establishes procedures to circumscribe presidential authority to use armed forces in potential or actual hostilities without congressional authorization.	R41199 R41989
12/31/1974	Privacy Act of 1974 (p. 30)	P.L. 93-579	88 Stat. 1896	5 U.S.C. §552[a]	Limits the disclosure of personally identifiable information (PII) held by federal agencies. Established a code of fair information practices for collection, management, and dissemination of records by agencies, including requirements for security and confidentiality of records.	
10/25/1978	Foreign Intelligence Surveillance Act of 1978 (FISA)	P.L. 95-511	92 Stat. 1783	18 U.S.C. §§2511, 2518-9, 50 U.S.C. Chapter 36, §§1801-1885c	In foreign intelligence investigations, provides a statutory framework for federal agencies to obtain authorization to conduct electronic surveillance, utilize pen registers and trap and trace devices, or access specified records.	98-326 R40138
10/13/1980	Privacy Protection Act of 1980	P.L. 96-440	94 Stat. 1879	42 U.S.C. Chapter 21A, §§2000aa-5 to 2000aa-12	Protects journalists from being required to turn over to law enforcement any work product and documentary materials, including sources, before dissemination to the public.	
10/12/1984	Counterfeit Access Device and Computer Fraud and Abuse	P.L. 98-473	98 Stat. 2190	18 U.S.C. §1030	Provided criminal penalties for unauthorized access and use of computers and networks. Part of the	97-1025

Table 2. (Continued)

Year	Popular Name	Law	Stat.	U.S.C.	Applicability and Notes	CRS Reports
	Act of 1984 (p. 31)				Comprehensive Crime Control Act of 1984.	
10/16/1986	Computer Fraud and Abuse Act of 1986	P.L. 99-474	100 Stat. 1213	18 U.S.C. §1030	Expanded the scope of the Counterfeit Access Device and Computer Fraud and Abuse Act of 1984. For government computers, criminalized electronic trespassing, exceeding authorized access, and destroying information; also criminalized trafficking in stolen computer passwords. Created a statutory exemption for intelligence and law enforcement activities.	
10/21/1986	Electronic Communications Privacy Act of 1986 (ECPA) (p. 32)	P.L. 99-508	100 Stat. 1848	18 U.S.C. §§2510-2522, 2701-2712, 3121-3126	Attempts to strike a balance between privacy rights and the needs of law enforcement with respect to data shared or stored by electronic and telecommunications services. Unless otherwise provided, prohibits the interception of or access to stored oral or electronic communications, use or disclosure of information so obtained, or possession of electronic eavesdropping equipment.	R41733 R41756 RL346
10/30/1986	Department of Defense Appropriations Act, 1987 (p. 34)	P.L. 99-591	100 Stat. 3341-82, 3341-122	10 U.S.C. §167	Established unified combatant command for special operations forces, including the U.S. Strategic Command, under which the U.S. Cyber	

Year	Popular Name	Law	Stat.	U.S.C.	Applicability and Notes	CRS Reports
					Command was organized.	
1/8/1988	Computer Security Act of 1987	P.L. 100-235	101 Stat. 1724	15 U.S.C. §§272, 278g-3, 278g-4, 278h	Required NIST to develop and the Secretary of Commerce to promulgate security standards and guidelines for federal computer systems except national security systems. Also required agency planning and training in computer security (this provision was superseded by FISMA—see p. 42).	
10/18/1988	Computer Matching and Privacy Protection Act of 1988	P.L. 100-503	102 Stat. 2507	5 U.S.C. §552[a]	Amended the Privacy Act (see p. 30), establishing procedural safeguards for use of computer matching on records covered by the act.	
12/9/1991	High Performance Computing Act of 1991 (p. 34)	P.L. 102-194	105 Stat. 1594	15 U.S.C. Chapter 81	Established a federal high-performance computing program and requires that it address security needs and provide for interagency coordination.	RL335
10/25/1994	Communications Assistance for Law Enforcement Act (CALEA) of 1994 (p. 36)	P.L. 103-414	108 Stat. 4279	47 U.S.C. §1001 et seq.	Requires telecommunications carriers to assist law enforcement in performing electronic surveillance and directs the telecommunications industry to design, develop, and deploy solutions that meet requirements for carriers to support authorized electronic surveillance.	RL306
5/25/1995	Paperwork Reduction Act of 1995	P.L. 104-13	109 Stat. 163	44 U.S.C. Chapter 35, §§3501-3549	Gave the Office of Management and Budget (OMB) authority to develop information-resource	

Table 2. (Continued)

Year	Popular Name	Law	Stat.	U.S.C.	Applicability and Notes	CRS Reports
					management polices and standards, required consultation with NIST and GSA on information technology (IT), and required agencies to implement processes relating to information security and privacy.	
2/8/1996	Telecommunications Act of 1996	P.L. 104-104	110 Stat. 56	See 47 U.S.C. §609 nt. for affected provisions.	Overhauled telecommunications law, including significant deregulation of U.S. telecommunications markets, eliminating regulatory barriers to competition.	
2/8/1996	Communications Decency Act of 1996 (p. 37)	P.L. 104-104 (Title V)	110 Stat. 133	See 47 U.S.C. §§223, 230	Intended to regulate indecency and obscenity on telecommunications systems, including the Internet. Has been interpreted to absolve Internet service providers and certain web-based services of responsibility for third-party content residing on those networks or websites.	R41499
2/10/1996	Clinger-Cohen Act (Information Technology Management Reform Act) of 1996) (p. 38)	P.L. 104-106, (Div. D and E)	110 Stat. 642	40 U.S.C. §11001 et seq.	Required agencies to ensure adequacy of information-security policies, OMB to oversee major IT acquisitions, and the Secretary of Commerce to promulgate compulsory federal computer standards based on those developed by NIST. Exempted national	

Year	Popular Name	Law	Stat.	U.S.C.	Applicability and Notes	CRS Reports
					security systems from most provisions.	
8/21/1996	Health Insurance Portability and Accountability Act of 1996 (HIPAA)	P.L. 104-191	110 Stat. 1936	42 U.S.C. §1320d et seq.	Required the Secretary of Health and Human Services to establish security standards and regulations for protecting the privacy of individually identifiable health information, and required covered health-care entities to protect the security of such information.	RL34120
10/11/1996	Economic Espionage Act of 1996	P.L. 104-294	110 Stat. 3488	18 U.S.C. §1030, Chapter 90, §§1831-1839	Outlaws theft of trade secret information, including electronically stored information, if "reasonable measures" have been taken to keep it secret. Also contains the National Information Infrastructure Protection Act of 1996, amending 18 U.S.C. §1030 (see the Counterfeit Access Device and Computer Fraud and Abuse Act of 1984, p. 31), broadening prohibited activities relating to unauthorized access to computers.	
10/30/1998	Identity Theft and Assumption Deterrence Act of 1998 (p. 39)	P.L. 105-318	112 Stat. 3007	18 U.S.C. §1028	Made identity theft a federal crime, provides penalties, and directed the FTC to record and refer complaints.	R40599
10/5/1999	National Defense Authorization Act for Fiscal Year 2000	P.L. 106-65	113 Stat. 512	10 U.S.C. §2224	Established the Defense Information Assurance Program and required development of a testbed and coordination with other federal agencies.	

Table 2. (Continued)

Year	Popular Name	Law	Stat.	U.S.C.	Applicability and Notes	CRS Reports
11/12/1999	Gramm-Leach-Bliley Act of 1999	P.L. 106-102 (Title V)	113 Stat. 1338	15 U.S.C. Chapter 94, §§6801-6827	Requires financial institutions to protect the security and confidentiality of customers' personal information; authorized regulations for that purpose.	RL34120 RS20185
10/30/2000	Floyd D. Spence National Defense Authorization Act for Fiscal Year 2001	P.L. 106-398 (Titles IX & X)	114 STAT. 1654A–233; 1654A–266	10 U.S.C. Chapter 112, §§2200-2200f	Established the DOD information assurance scholarship program; set cybersecurity requirements for federal systems superseded by FISMA in 2002	
10/26/2001	USA PATRIOT Act of 2001	P.L. 107-56	115 Stat. 272	see 18 U.S.C. §1 nt. and classification tables.a	Authorized various law-enforcement activities relating to computer fraud and abuse.	R40980
7/30/2002	Sarbanes-Oxley Act of 2002	P.L. 107-204	116 Stat. 745	15 U.S.C. §7262	Requires annual reporting on internal financial controls of covered firms to the Securities and Exchange Commission (SEC). Such controls typically include information security.	
11/25/2002	Homeland Security Act of 2002 (HSA) (p. 40)	P.L. 107-296 (Titles II and III)	116 Stat. 2135	6 U.S.C. §§121-195c, 441-444, and 481-486	Created the Department of Homeland Security (DHS) and gave it functions relating to the protection of information infrastructure, including providing state and local governments and private entities with threat and vulnerability information, crisis-management support, and technical assistance. Strengthened some criminal penalties	

Year	Popular Name	Law	Stat.	U.S.C.	Applicability and Notes	CRS Reports
					relating to cybercrime.	
11/25/2002	Federal Information Security Management Act of 2002 (FISMA) (p. 42)	P.L. 107-296 (Title X) P.L. 107-347 (Title III)	116 Stat. 2259 116 Stat. 2946	44 U.S.C. Chapter 35, Subchapters II and III, 40 U.S.C. 11331, 15 U.S.C. 278g-3 & 4	Created a cybersecurity framework for federal information systems, with an emphasis on risk management, and required implementation of agency-wide information security programs. Gave oversight responsibility to OMB, revised the responsibilities of the Secretary of Commerce and NIST for information-system standards, and transferred responsibility for promulgation of those standards from the Secretary of Commerce to OMB.	
11/26/2002	Terrorism Risk Insurance Act of 2002 (p. 45)	P.L. 107-297	116 Stat. 2322	15 U.S.C. §6701 nt.	Provides federal cost-sharing subsidies for insured losses resulting from acts of terrorism.	
11/27/2002	Cyber Security Research and Development Act, 2002 (p. 45)	P.L. 107-305	116 Stat. 2367	15 U.S.C. §§278g, h, 7401 et seq.	Requires the National Science Foundation (NSF) to award grants for basic research and education to enhance computer security. Required NIST to establish cybersecurity research programs.	
12/17/2002	E-Government Act of 2002 (p. 47)	P.L. 107-347	116 Stat. 2899	5 U.S.C. Chapter 37, 44 U.S.C. §3501 nt., Chapter 35, Subchapter 2, and Chapter 36	Serves as the primary legislative vehicle to guide federal IT management and initiatives to make information and services available online. Established the Office of Electronic Government within OMB, the Chief Information Officers	

Table 2. (Continued)

Year	Popular Name	Law	Stat.	U.S.C.	Applicability and Notes	CRS Reports
					(CIO) Council, and a government/private-sector personnel exchange program; includes FISMA; established and contains various other requirements for security and protection of confidential information.	
12/4/2003	Fair and Accurate Credit Transactions Act of 2003	P.L. 108-159	117 Stat. 1952	See 15 U.S.C. §1601 nt. for affected provisions.	Required the FTC and other agencies to develop guidelines for identity theft prevention programs in financial institutions, including "red flags" indicating possible identity theft.	RS20185
12/16/2003	Controlling the Assault of Non-Solicited Pornography and Marketing (CAN-SPAM) Act of 2003	P.L. 108-187	117 Stat. 2699	15 U.S.C. Chapter 103, §§7701-7713, 18 U.S.C. 1037	Imposed regulations on the transmission of unsolicited commercial email, including prohibitions against predatory and abusive email, and false or misleading transmission of information.	
7/15/2004	Identity Theft Penalty Enhancement Act 2004 (p. 48)	P.L. 108-275	118 Stat. 831	18 U.S.C. §§1028, 1028A	Established penalties for aggravated identity theft.	R40599
12/17/2004	Intelligence Reform and Terrorism Prevention Act of 2004 (IRPTA) (p. 49)	P.L. 108-458	118 Stat. 3638	42 U.S. C. §2000ee, 50 U.S.C. §403-1 et seq., §403-3 et seq., §404o et. seq.	Created the position of Director of National Intelligence (DNI). Established mission responsibilities for some entities in the intelligence, homeland security, and national security communities, and established a Privacy and Civil Liberties Board within the Executive Office of the President.	

Year	Popular Name	Law	Stat.	U.S.C.	Applicability and Notes	CRS Reports
8/8/2005	Energy Policy Act of 2005 (EPACT)	P.L. 109-58	119 Stat. 594	16 U.S.C. 824o	Requires FERC to certify an Electric Reliability Organization (ERO) to establish and enforce reliability standards for bulk electric-power system facilities.	R41886
10/4/2006	Department of Homeland Security Appropriations Act, 2007	P.L. 109-295	120 Stat. 1355	6 U.S.C. §121 nt.	Sec. 550 required the Secretary of Homeland Security to issue regulations (6 C.F.R. Part 27) establishing risk-based performance standards for security of chemical facilities; regulations include cybersecurity standards requirement (6 C.F.R. §27.230(a)(8)).	
8/5/2007	Protect America Act of 2007	P.L. 110-55	121 Stat. 552	50 U.S.C. §1801 nt.	Provided authority for the Attorney General and the DNI to gather foreign intelligence information on persons believed to be overseas. The act expired in 2008.	
12/19/2007	Energy Independence and Security Act of 2007 (EISA)	P.L. 110-140	121 Stat. 1492	42 U.S.C. §§17381-17385	Gave NIST primary responsibility for developing interoperability standards for the electric-power "smart grid."	R41886
7/10/2008	Foreign Intelligence Surveillance Act of 1978 [FISA] Amendments Act of 2008	P.L. 110-261 P.L.	122 Stat. 2436	See 50 U.S.C. §1801 nt. for affected provisions.	Added additional procedures to FISA (see p. 54) for acquisition of communications of persons outside the United States.	98-326
9/26/2008	Identity Theft Enforcement and Restitution Act of 2008	110-326	122 Stat. 356	18 U.S.C. §1030	Authorized restitution to identity theft victims and modified some of the activities and penalties covered by 18 U.S.C. 1030.	R40599 97-1025

Table 2. (Continued)

Year	Popular Name	Law	Stat.	U.S.C.	Applicability and Notes	CRS Reports
2/17/2009	Health Information Technology for Economic and Clinical Health Act	P.L. 111-5 (Title XIII of Div. A and Title IV of Div. B)	123 Stat. 226	42 U.S.C. §17901 et seq.	Expanded privacy and security requirements for protected health information by broadening HIPAA breach disclosure notification and privacy requirements to include business associates of covered entities.	R40546

Source: Various sources (see text), including National Research Council, Toward a Safer and More Secure Cyberspace (Washington, DC: National Academy Press, 2007); The White House, Cyberspace Policy Review, May 29, 2009, http://www.whitehouse.gov/assets and CRS.

Note: Prepared by Rita Tehan, Information Research Specialist (rtehan@crs.loc.gov, 7-6739) and Eric A. Fischer. Laws in italics are discussed in the text.

[a] Office of the Law Revision Counsel, "United States Code Table of Classifications for Public Laws, 107th Congress, 1st Session (Covering Public Laws 107-1 through 107-136)," http://uscode.house.gov/classification tbl107pl_1st.htm.

End Notes

[1] The term *information systems* is defined in 44 U.S.C. §3502 as "a discrete set of information resources organized for the collection, processing, maintenance, use, sharing, dissemination, or disposition of information," where *information resources* is "information and related resources, such as personnel, equipment, funds, and information technology." Thus *cybersecurity,* a broad and arguably somewhat fuzzy concept for which there is no consensus definition, might best be described as measures intended to protect information systems—including technology (such as devices, networks, and software), information, and associated personnel—from various forms of attack. The concept has, however, been characterized in various ways. For example, the interagency Committee on National Security Systems has defined it as "the ability to protect or defend the use of cyberspace from cyber attacks," where *cyberspace* is defined as "a global domain within the information environment consisting of the interdependent network of information systems infrastructures including the Internet, telecommunications networks, computer systems, and embedded processors and controllers" (Committee on National Security Systems, *National Information Assurance (IA) Glossary*, April 2010, http://www.cnss.gov/Assets/pdf/cnssi_4009.pdf). In contrast, cybersecurity has also been defined as synonymous with *information security* (see, for example, S. 773, the Cybersecurity Act of 2010, in the 111th Congress), which is defined in current law (44 U.S.C. §3532(b)(1)) as protecting information and information systems from unauthorized access, use, disclosure, disruption,

modification, or destruction in order to provide—integrity, which means guarding against improper information modification or destruction, and includes ensuring information nonrepudiation and authenticity; confidentiality, which means preserving authorized restrictions on access and disclosure, including means for protecting personal privacy and proprietary information; availability, which means ensuring timely and reliable access to and use of information; and authentication, which means utilizing digital credentials to assure the identity of users and validate their access.

[2] See, for example, IBM, *IBM X-Force® 2011 Mid-year Trend and Risk Report*, September 2011, http://public.dhe.ibm.com/common/ssi/ecm/en/wgl03009usen/ WGL03009USEN.PDF; Barbara Kay and Paula Greve, *Mapping the Mal Web IV* (McAfee, September 28, 2010), http://us.mcafee.com/en-us/local/docs/MTMW_Report.pdf; Office of the National Counterintelligence Executive, *Foreign Spies Stealing U.S. Economic Secrets in Cyberspace: Report to Congress on Foreign Economic Collection and Industrial Espionage, 2009-2011*, October 2011, http://www.ncix.gov/publications/reports/ fecie_all/ Foreign_Economic_Collection_2011.pdf; Symantec, *Symantec Internet Security Threat Report: Trends for 2010*, Volume 16, April 2011, https://www4.symantec.com/mktginfo/ downloads/21182883_GA_REPORT_ISTR_Main-Report_04-11_HI-RES.pdf.

[3] This term is defined in 44 U.S.C. §3542(b)(2).

[4] Those bills were identified through a two-step process—candidates were found through searches of the Legislative Information System (LIS, http://www.congress) using "cybersecurity," "information systems," and other relevant terms in the text of the bills, followed by examination of that text in the candidates to determine relevance for cybersecurity. Use of other criteria may lead to somewhat different results. For example, using the LIS "cybersecurity" topic search yields about 30 bills in the 112th Congress and 40 in the 111th, with about a 50% overlap in the bills included. While that difference is higher than might be expected, none of the bills identified uniquely by the LIS topic search are relevant to the discussion in this report.

[5] Among the broader proposals in the 111th Congress, S. 773 (S.Rept. 111-384) and S. 3480 (S.Rept. 111-368) were reported by the originating committees. H.R. 4061 (H.Rept. 111-405) and H.R. 5136 (Title XVII, mostly similar to H.R. 4900) both passed the House. A bill combining provisions of the two Senate bills was drafted (Tony Romm, "Lack of Direction Slows Cybersecurity," *Politico*, November 4, 2010, http://www.politico.com/news/ stories/1110/ 44662.html). In the 112th Congress, S. 413 is similar to S. 3480 in the previous Congress, H.R. 2096 (H.Rept. 112-264) is similar to H.R. 4061, and the Senate combined bill, S. 2105, includes elements of S. 773, S. 413, S. 2102, and a proposal put forward by the White House in April 2011 (see below).

[6] This update does not include executive branch actions taken since December 2011.

[7] The White House, "The Comprehensive National Cybersecurity Initiative," March 5, 2010, http://www.whitehouse.gov/cybersecurity/comprehensive-national-cybersecurity-initiative. For additional information about this initiative and associated policy considerations, see CRS Report R40427, *Comprehensive National Cybersecurity Initiative: Legal Authorities and Policy Considerations*, by John Rollins and Anna C. Henning.

[8] The position has been popularly called the "cyber czar."

[9] Jeffrey Zients, Vivek Kundra, and Howard A. Schmidt, "FY 2010 Reporting Instructions for the Federal Information Security Management Act and Agency Privacy Management," Office of Management and Budget, Memorandum for Heads of Executive Departments and Agencies M-10-15, April 21, 2010, http://www.whitehouse.gov/omb/assets/ memoranda_ 2010/m 10-15.pdf.

[10] See, for example, Seymour M. Hersh, "Judging the cyber war terrorist threat," *The New Yorker*, November 1, 2010, http://www.newyorker.com/reporting/2010/11/01/ 101101 fa_fact_hersh?currentPage=all.

[11] Among them are White House strategies to improve the security of Internet transactions (The White House, *National Strategy for Trusted Identities in Cyberspace*, April 2011, http://www.whitehouse.gov/sites/default/files/rss_viewer/ NSTICstrategy_041511.pdf) and to coordinate international efforts (The White House, *International Strategy for Cyberspace*, May 2011, http://www.whitehouse.gov/sites/default/files/rss_viewer/ international_strategy_for_cyberspace.pdf), and an executive order on sharing and security for classified information (Executive Order 13587, "Structural Reforms to Improve the Security of Classified Networks and the Responsible Sharing and Safeguarding of Classified Information," *Federal Register* 76, no. 198 (October 13, 2011): 63811-63815, http://www.gpo.gov/fdsys/pkg/FR-2011-10-13/pdf/2011-26729.pdf).

[12] This responsibility is shared to some extent with other agencies such as the U.S. Secret Service.

[13] For specific analysis of legal issues associated with several of the bills being debated in the 112th Congress, see CRS Report R42409, *Cybersecurity: Selected Legal Issues*, by Edward C. Liu et al.

[14] The title on information sharing is similar to S. 2102.

[15] SECURE IT is an acronym for Strengthening and Enhancing Cybersecurity by Using Research, Education, Information and Technology.

[16] A very similar but not identical bill, H.R. 4263, was introduced in the House April 9. It is not discussed separately in this update.

[17] The White House, *Complete Cybersecurity Proposal*, 2011, http://www.whitehouse.gov/ sites/default/files/omb/ legislative/letters/law-enforcement-provisions-related-to-computer-security-full-bill.pdf. One part does not appear to be directly related to cybersecurity. It would restrict the authority of state and local jurisdictions with respect to the location of commercial data centers.

[18] House Republican Cybersecurity Task Force, *Recommendations of the House Republican Cybersecurity Task Force*, October 5, 2011, http://thornberry.house.gov/UploadedFiles/ CSTF_Final_Recommendations.pdf.

[19] The Obama Administration has objected to this bill, claiming that it does not address cybersecurity needs for critical infrastructure, and contains overly broad liability protections for private-sector entities and insufficient protections for individual privacy, confidentiality, and civil liberties (The White House, "H.R. 3523—Cyber Intelligence Sharing and Protection Act," Statement of Administration Policy, April 25, 2012, http://www. whitehouse.gov/sites/default/files/ omb/legislative/sap/112/saphr3523r_20120425.pdf). The Administration has not released statements of administration policy for any of the other bills discussed in this report.

[20] For discussion of this bill and H.R. 2096, see also CRS Report RL33586, *The Federal Networking and Information Technology Research and Development Program: Background, Funding, and Activities*, by Patricia Moloney Figliola.

[21] H.R. 3674 was marked up by the Subcommittee on Cybersecurity, Infrastructure Protection, and Security Technologies of the Committee on Homeland Security on February 1 and forwarded to the full committee, which substantially amended the bill in its April 18 markup.

[22] See also "Federal Information Security Management Act of 2002 (FISMA)."

[23] See CRS Report R42409, *Cybersecurity: Selected Legal Issues* for more detail.

[24] As used here, *civil systems* means federal information systems other than national security systems (defined in 44 U.S.C. §3542) and mission-critical Department of Defense and Intelligence Community systems (i.e., compromise of those systems "would have a debilitating impact on the mission" of the agencies [see 44 U.S.C. 3543(c)]).

[25] See, for example, CSIS Commission on Cybersecurity for the 44th Presidency, *Securing Cyberspace for the 44th Presidency*, December 2008, http://www.csis.org/tech/cyber/; Partnership for Public Service and Booz Allen Hamilton, *Cyber IN-Security: Strengthening the Federal Cybersecurity Workforce*, July 2009, http://ourpublicservice.org/OPS/ publications/download.php?id=135; CSIS Commission on Cybersecurity for the 44th Presidency, *A Human Capital Crisis in Cybersecurity*, July 2010, http://csis.org/files /publication/ 100720_Lewis_HumanCapital_WEB_BlkWhteVersion.pdf.

[26] This includes providing requirements or statutory authority for existing programs, such as the joint NSF/DHS Scholarship-for Service Program (see Office of Personnel Management, "Federal Cyber Service: Scholarship For Service," n.d., https://www.sfs.opm.gov/; National Science Foundation, *Federal Cyber Service: Scholarship for Service (SFS)*, NSF 08-600, Program Solicitation, December 2, 2008, http://www.nsf.gov/pubs/2008/ns f08600/nsf08600.htm), the NSA/DHS National Centers of Academic Excellence and National Security Agency ("National Centers of Academic Excellence," January 10, 2012, http://www.nsa.gov/ia/academic_outreach/nat_cae/index.shtml), and the U.S. Cyber Challenge (National Board of Information Security Examiners, "US Cyber Challenge," 2012, https://www.nbise.org/uscc).

[27] See, for example, National Research Council, *Trust in Cyberspace* (Washington, DC: National Academies Press, 1999), http://www.nap.edu/catalog/6161.html.

[28] The percentages were calculated from data in Subcommittee on Networking and Information Technology Research and Development, Committee on Technology, *Supplement to the President's Budget for Fiscal Year 2013: The Networking and Information Technology Research and Development Program*, February 2012, http://www.nitrd.gov/ PUBS% 5C2013supplement%5CFY13NITRDSupplement.pdf. The total investment for FY2011 was $445 million. However, agencies may perform additional research not reported as cybersecurity R&D (e.g., some research on software design or high-confidence systems).

[29] For example, through the Director of the Office of Science and Technology Policy (OSTP).

[30] This authority had been granted to the Secretary of Commerce under the Clinger-Cohen Act of 1996 (P.L. 104-106) but was transferred to the Director of OMB by the FISMA title in the HSA in 2002 (P.L. 107-296, Sec. 1002, 40 U.S.C. §11331). Note that the version of the Chapter 35 provisions that is currently in effect (Subchapter III) was enacted by the FISMA title in the E-Government Act of 2002 (P.L. 107-347, Title III), but that is not the case for 40 U.S.C. §11331, for which the version in the E-Government Act would have retained the authority of the Secretary of Commerce to promulgate those standards, even though it was enacted after the HSA. The reason for this potentially confusing difference appears to be that (1) the effective date of HSA was later than that of the E-Government Act, and (2) HSA changed 44 U.S.C. Chapter 35 by amending the existing subchapter II, which the E-Government Act explicitly suspended (see also "Federal Information Security Management Act of 2002 (FISMA)").

[31] See Jeffrey Zients, Vivek Kundra, and Howard A. Schmidt, "FY 2010 Reporting Instructions for the Federal Information Security Management Act and Agency Privacy Management," Office of Management and Budget, Memorandum for Heads of Executive Departments and Agencies M-10-15, April 21, 2010, http://www.whitehouse.gov/omb/assets/ memoranda_ 2010/m10-15.pdf; and Peter R. Orszag and Howard A. Schmidt, "Clarifying Cybersecurity

Responsibilities and Activities of the Executive Office of the President and the Department of Homeland Security (DHS)," Office of Management and Budget, Memorandum for Heads of Executive Departments and Agencies M-10-28, July 6, 2010, http://www.whitehouse.gov/sites/default/files/omb/assets/memoranda_2010/m10- 28.pdf.

[32] See Department of Homeland Security, "Critical Infrastructure", May 4, 2012, http://www.dhs.gov/files/programs/ gc_1189168948944.shtm; and CRS Report RL30153, *Critical Infrastructures: Background, Policy, and Implementation*, by John D. Moteff.

[33] See, for example, House Committee on Homeland Security, Subcommittee on Cybersecurity, Infrastructure Protection, and Security Technologies, *Examining the Cyber Threat to Critical Infrastructure and the American Economy*, 2011, http://homeland.house.gov/hearing/subcommittee-hearing-examining-cyber-threat-criticalinfrastructure-and-american-economy; Stewart Baker, Natalia Filipiak, and Katrina Timlin, *In the Dark: Crucial Industries Confront Cyberattacks* (McAfee and CSIS, April 21, 2011), http://www.mcafee.com/us/resources/reports/rpcritical-infrastructure-protection.pdf; and R. E. Kahn et al., *America's Cyber Future: America's Cyber Future: Security and Prosperity in the Information Age* (Center for a New American Security, May 31, 2011), http://www.cnas.org/files/ documents/publications/CNAS_Cyber_Volume%20I_0.pdf.

[34] S. 2105 would largely exempt information technology products and services from designation as covered CI and the cybersecurity regulations the bill would authorize.

[35] An entity would be exempted if the Secretary of Homeland Security determined that it was already sufficiently secure or that additional requirements would not substantially improve its security (Sec. 105(c)(4)). The President would also be permitted to exempt an entity from the requirements upon determining that current regulations sufficiently mitigate the risks to the entity (Sec. 104(f)).

[36] This exemption (Sec. 9(c) in the part of the proposal on CI protection) is similar to the Presidential exemption in S. 2105 (footnote 35) except that the *White House Proposal* would give the authority to the Secretary of Homeland Security.

[37] Among the possibilities discussed are tying adoption of standards to incentives such as grants and streamlined regulation, using tax credits, and facilitating the development of a cybersecurity insurance market.

[38] This is the version approved by voice vote by the Subcommittee on Cybersecurity, Infrastructure Protection, and Security Technologies of the House Committee on Homeland Security on February 1, 2012, and forwarded to the full committee.

[39] See Department of Homeland Security, *National Infrastructure Protection Plan*, 2009, http://www.dhs.gov/xlibrary/ assets/NIPP_Plan.pdf.

[40] The criteria in the subcommittee version of H.R. 3674 are generally similar to those in S. 2105 and the *White House Proposal* in that they focus on entities for which successful cyberattack could have major negative impacts. The definitions in the three legislative proposals differ somewhat in emphasis and specificity.

[41] This is the version ordered reported by the Committee on Homeland Security on April 18, 2012.

[42] See, for example, The Markle Foundation Task Force on National Security in the Information Age, *Nation At Risk: Policy Makers Need Better Information to Protect the Country*, March 2009, http://www.markle.org/ downloadable_assets/20090304_mtfjeport.pdf; CSIS Commission on Cybersecurity for the 44th Presidency, *Cybersecurity Two Years Later*, January 2011, http://csis.org/files/publication/ 110128_Lewis_CybersecurityTwoYearsLater_Web.pdf.

[43] See, for example, Greg Nojeim, "WH Cybersecurity Proposal: Questioning the DHS Collection Center," *Center for Democracy & Technology*, May 24, 2011, http://cdt.org /blogs/greg-nojeim/wh-cybersecurity-proposal-questioning-dhscollection-center; and Adriane Lapointe, *Oversight for Cybersecurity Activities* (Center for Strategic and International Studies, December 7, 2010), http://csis.org/files/publication/ 101202_Oversight_for_Cybersecurity_Activities.pdf. See also comments received by a Department of Commerce task force (available at http://www.nist.gov/itl/ cybersecnoi.cfm) in conjunction with development of this report: Internet Policy Task Force, *Cybersecurity, Innovation, and the Internet Economy* (Department of Commerce, June 2011), http://www.nist.gov/itl/upload/ Cybersecurity_Green-Paper_FinalVersion.pdf. See also footnote 19.

[44] House Committee on Homeland Security, Subcommittee on Cybersecurity, Infrastructure Protection, and Security Technologies, "Hearing on Draft Legislative Proposal on Cybersecurity," 2011, http://homeland.house.gov/hearing/ subcommittee-hearing-hearing-draft-legislative-proposal-cybersecurity.

[45] Department of Homeland Security, "National Cybersecurity and Communications Integration Center", December 6, 2011, http://www.dhs.gov/files/programs/nccic.shtm.

[46] Department of Homeland Security Office of Inspector General, "Secretary Napolitano Opens New National Cybersecurity and Communications Integration Center," Press Release, October 30, 2009, http://www.dhs.gov/ynews/ releases/pr_1256914923094.shtm. The subcommittee version of H.R. 3476 would also have provided statutory authority for NCCIC, but would have given it somewhat different responsibilities.

[47] The committee version of H.R. 3674 includes a FOIA exemption by reference to the amendments to Title XI of the "National Security Act of 1947" that would be made by H.R. 3523.

[48] Sec. 1104(c)(7) of the National Security Act as added by Sec. 2(a) of the bill.

[49] Sec. 1104(b)(3)(A) as added.

[50] The board was established by the "Intelligence Reform and Terrorism Prevention Act of 2004 (IRTPA)."

[51] Sec. 245(a)(1) as added to the HSA by the proposal.

[52] For discussion of federal cybercrime laws, see CRS Report 97-1025, *Cybercrime: An Overview of the Federal Computer Fraud and Abuse Statute and Related Federal Criminal Laws*, by Charles Doyle; and CRS Report R40599, *Identity Theft: Trends and Issues*, by Kristin M. Finklea. See also the discussions of criminal statutes in this report.

[53] There are 27 entries, but the one on antitrust laws consists of four different statutes. Neither of the two lists is intended to be definitive or exhaustive. For example, some analysts may argue that more agency authorization statutes should be included, or, alternatively, that some of the statutes that are included are not of significant relevance.

[54] This is the name by which the statute is commonly known.

[55] The public law (P.L.) and *United States Statutes at Large* (Stat.) citations refer to the original law to which the popular name currently applies. Laws enacted before 1957 generally do not have public law numbers but chapter numbers (Ch.) instead. U.S. Code (U.S.C.) citations refer to the codified law, including any amendments, of those provisions deemed most relevant for cybersecurity as discussed in the text under that law (see also footnote 56). For more information about citation forms, see Law Library of Congress, "Federal Statutes," April 4, 2011, http://www.loc.gov/law/help/statutes.php. More complete cross-references of public laws to corresponding provisions of U.S. Code can be found in classification tables (see, for example, U.S. House of Representatives, Office of the Law

Revision Counsel, "U.S. Code Classification Tables," 2011, http://uscode.house.gov/ classification/tables.shtml).

[56] In some cases, such as the Cybersecurity Research And Development Act, P.L. 107-305, the entire statute is relevant to cybersecurity. In others, such as the Omnibus Crime Control and Safe Streets Act of 1968, P.L. 90-351, the statute has a broader focus and only the provisions relevant to the text are cited and described. However, given that cybersecurity is not a precise concept, there may in some cases be legitimate disagreements among experts about which provisions are relevant. Therefore, the descriptions and U.S. Code citations cannot be considered definitive.

[57] The discussion is provided for purposes of information only. CRS does not propose legislation or take positions or make recommendations on legislative proposals or issues. Contributing CRS staff include Patricia Moloney Figliola, Kristin M. Finklea, Eric A. Fischer, Wendy R. Ginsberg, John Rollins, Kathleen Ann Ruane, Gina Stevens, Rita Tehan, and Catherine A. Theohary. Entries for which no contributor is indicated were written by Eric A. Fischer.

[58] The order is by date of enactment of the earliest relevant statute, as assessed by CRS. This organization, rather than alternatives such as by topic or U.S. Code title, was chosen because it provides the best view of the evolution of legislation in this area.

[59] Sources are cited where they could be specifically identified.

[60] Data-breach notification is also covered by H.R. 1528, H.R. 1707, H.R. 1841, H.R. 2577, S. 1151, S. 1207, S. 1480, and S. 1535.

[61] This act was classified to 15 titles.

[62] Prepared by Catherine A. Theohary, Analyst in National Security Policy and Information Operations (ctheohary@crs.loc.gov, 7-0844).

[63] For further discussion, see CRS Report RS22266, *The Use of Federal Troops for Disaster Assistance: Legal Issues*, by Jennifer K. Elsea and R. Chuck Mason.

[64] For example, see Jeffrey K. Toomer, "A Strategic View of Homeland Security: Relooking the Posse Comitatus Act and DOD's Role in Homeland Security" (monograph, School of Advanced Military Studies, United States Army Command and General Staff College, Fort Leavenworth, Kansas, July 11, 2002), http://www.dtic.mil/cgi-bin/ GetTRDoc?Location= U2&doc=GetTRDoc.pdf&AD=ADA403866.

[65] Department of Homeland Security and Department of Defense, "Regarding Cybersecurity." The MOA provides terms for sharing of personnel, equipment, and facilities by the two agencies to improve planning, capabilities, and mission activities in national cybersecurity efforts.

[66] Prepared by Kathleen Ann Ruane, Legislative Attorney (kruane@crs.loc.gov, 7-9135).

[67] See, e.g., United States v. American Airlines Inc., 743 F.2d 1114 (5th Cir. 1984); FTC v. Motion Picture Advertising Serv. Co., 344 U.S. 392, 394-95 (1953); FTC v. Cement Institute, 333 U.S. 683, 694 (1948); Fashion Originators' Guild v. FTC, 312 U.S. 457, 463-64 (1941).

[68] See Standard Oil Co. v. U.S., 221 U.S. 1 (1911).

[69] See Federal Trade Commission and Department of Justice, *Antitrust Guidelines for Collaborations among Competitors*, April 2000, http://www.ftc.gov/os/2000/04/ftcdo jguidelines.pdf.

[70] Ibid.

[71] Ibid. (noting that many collaborations among competitors are "not only benign, but procompetitive").

[72] House Republican Cybersecurity Task Force, *Recommendations*, p. 11.

[73] See Amitai Aviram, "Network Responses to Network Threats," in *The Law and Economics of Cybersecurity*, ed. Mark Grady and Francesco Parisi (New York: Cambridge University Press, 2006), 157-158.

[74] See Federal Trade Commission and Department of Justice, *Antitrust Guidelines*.

[75] Ibid.

[76] 28 C.F.R. §50.6.

[77] Federal Trade Commission and Department of Justice, *Antitrust Guidelines*.

[78] Joel I. Klein, Assistant Attorney General, to Barbara Greenspan, Associate General Counsel, Electric Power Institute, Inc., October 2, 2000, http://www.justice.gov/atr/public/busreview/6614.htm.

[79] 15 U.S.C. §272, as amended by the Technology Competitiveness Act, Subtitle B of Title V of P.L. 100-418, the Omnibus Trade and Competitiveness Act of 1988, which also changed the name of the agency from the National Bureau of Standards to the National Institute of Standards and Technology, and changed the name of the act to the National Institute of Standards and Technology Act.

[80] 15 U.S.C. §§278g-3 and -4, as added by the Computer Security Act of 1987. See also "Federal Information Security Management Act of 2002 (FISMA)."

[81] The law was originally enacted in 1920 as the Federal Water Power Act but was renamed the Federal Power Act in 1935 (49 Stat. 863, 16 U.S.C. §791a).

[82] See, for example, H.Rept. 111-493, S.Rept. 111-331.

[83] CRS Report R41886, *The Smart Grid and Cybersecurity—Regulatory Policy and Issues*, by Richard J. Campbell.

[84] See also "Communications Decency Act of 1996."

[85] See, for example, Elgin M. Brunner and Manuel Suter, *International CIIP Handbook 2008/2009* (Center for Security Studies, ETH Zurich, 2008), http://www.css.ethz.ch/publications/CIIP_HB_08.

[86] See also CRS Report R41674, *Terrorist Use of the Internet: Information Operations in Cyberspace*, by Catherine A. Theohary and John Rollins.

[87] S. 413 is largely identical to S. 3480. Both would provide the authority for the emergency measures through a revision of the Homeland Security Act, not the Communications Act. In addition, they would assign the authority to implement Sec. 706 to the head of a White House office to be created by the bills. The provision in S. 773 was not presented as a revision to a specified law.

[88] For example, the *Task Force Report* states, "There is widespread agreement that greater sharing of information is needed within industries, among industries, and between government and industry in order to improve cybersecurity and to prevent and respond to rapidly changing threats. For example, through intelligence collection, the federal government has insights and capabilities that many times are classified but would be useful to help defend private companies from cybersecurity attacks" (House Republican Cybersecurity Task Force, *Recommendations*, p. 10).

[89] Prepared by Catherine A. Theohary, Analyst in National Security Policy and Information Operations (ctheohary@crs.loc.gov, 7-0844).

[90] This restriction was added by the Foreign Relations Authorization Act, Fiscal Years 1986 and 1987 (P.L. 99-93, 99 Stat. 431) and was not part of the original act.

[91] For discussion, see CRS Report R40989, *U.S. Public Diplomacy: Background and Current Issues*, by Kennon H. Nakamura and Matthew C. Weed.

[92] See, for example, CSIS Commission on Cybersecurity for the 44th Presidency, *Securing Cyberspace for the 44th Presidency*, December 2008, http://www.csis.org/tech/cyber/; The

White House, *Cyberspace Policy Review*, May 29, 2009, http://www.whitehouse.gov/assets/documents/Cyberspace_Policy_Review_final.pdf; and The White House, *International Strategy for Cyberspace*.

[93] Prepared by Wendy R. Ginsberg, Analyst in Government Organization and Management (wginsberg@crs.loc.gov, 7- 3933).

[94] The statute must require that the data be withheld from the public in such a manner as to leave no discretion on the issue, establish particular criteria for withholding information or refer to particular types of matters to be withheld, or specifically cite the exemption if enacted after October 28, 2009, the date of enactment of the OPEN FOIA Act of 2009, P.L. 111-83. These exemptions are also called "b(3) exemptions" because they are created pursuant to 5 U.S.C. §552(b)(3).

[95] Other exemptions may also sometimes apply to cybersecurity information. For further discussion of FOIA and its exemptions, see CRS Report R41933, *Freedom of Information Act (FOIA): Background and Policy Options for the 112th Congress*, by Wendy Ginsberg, CRS Report R41406, *The Freedom of Information Act and Nondisclosure Provisions in Other Federal Laws* , by Gina Stevens.

[96] See "Sec. 245. Voluntary Disclosure of Cybersecurity Information," in The White House, "Department of Homeland Security Cybersecurity Authority and Information Sharing," May 12, 2011, p. 8–9, http://www.whitehouse.gov/sites/ default/files/omb/legislative/letters/dhs-cybersecurity-authority.pdf.

[97] Specifically, it states, "information sharing within existing structures can be improved through limited safe harbors when private sector entities voluntarily disclose threat, vulnerability, or incident information to the federal government or ask for advice or assistance to help increase protections on their own systems. These protections would need to address concerns about antitrust issues, liability, an exemption from the Freedom of Information Act (FOIA), protection from public disclosure, protection from regulatory use by government, and whether or not a private entity is operating as an agent of the government. However, the protection of personal privacy should be at the forefront of any limited legal protection proposal" (House Republican Cybersecurity Task Force, *Recommendations*, p. 11).

[98] These provisions, along with possible updates, are discussed under "Electronic Communications Privacy Act of 1986."

[99] There is no uniform definition of "cybercrime." Furthermore, no definitive statistics on cybercrime appear to be publically available. However, the public/private Internet Crime Complaint Center referred 25 times as many of the complaints it received to law enforcement agencies in 2010 (121,710) as in 2001 (4,810) (Internet Crime Complaint Center, *2010 Internet Crime Report*, 2011, http://www.ic3.gov/media/ annualreport/2010_IC3Report.pdf).

[100] For details, CRS Report 96-950, *RICO: A Brief Sketch*, by Charles Doyle.

[101] House Republican Cybersecurity Task Force, *Recommendations*, p. 14.

[102] For more information, see CRS Report R40520, *Federal Advisory Committees: An Overview*, by Wendy Ginsberg.

[103] Isabelle Abele-Wigert and Myriam Dunn, *International CIIP Handbook 2006, Vol. I* (Center for Security Studies, ETH Zurich, 2006), p. 337, http://www.css.ethz.ch/publications/CIIP_HB_06_Vol.1.pdf; Brunner and Suter, *International CIIP Handbook 2008/2009*, p. 456.

[104] House Committee on Homeland Security, "Hearing on Draft Legislative Proposal on Cybersecurity."

[105] For information on how they have been interpreted by the courts, see Department of Justice, "Overview of the Privacy Act of 1974, 2010 Edition," March 2, 2010, http://www.justice.gov/opcl/1974privacyact-overview.htm.

[106] The Computer Fraud and Abuse Act of 1986 (P.L. 99-474, 100 Stat. 1213) expanded the scope of the original act. For government computers, it criminalized electronic trespassing, exceeding authorized access, and destroying information. It also criminalized trafficking in stolen computer passwords and created a statutory exemption for intelligence and law enforcement activities.

[107] For more information, see CRS Report 97-1025, *Cybercrime: An Overview of the Federal Computer Fraud and Abuse Statute and Related Federal Criminal Laws*, by Charles Doyle.

[108] See CRS Report R41941, *The Obama Administration's Cybersecurity Proposal: Criminal Provisions*, by Gina Stevens.

[109] House Republican Cybersecurity Task Force, *Recommendations*, p. 14.

[110] Prepared by Gina Stevens, Legislative Attorney (gstevens@crs.loc.gov, 7-2581).

[111] 100 Stat. 1848; see also House Committee on the Judiciary, "Electronic Communications Privacy Act of 1986," H.Rept. 99-647, 99th Cong. 2d Sess. 2, at 19 (1986).

[112] House Committee on the Judiciary, Subcommittee on the Constitution, Civil Rights, and Civil Liberties, *ECPA Reform and the Revolution in Cloud Computing*, 2010, http://judiciary.house.gov/hearings/hear_100923.html (statement of Edward W. Felton, Professor Princeton University): In 1986, when ECPA was passed, the Internet consisted of a few thousand computers. The network was run by the U.S. government for research and education purposes, and commercial activity was forbidden. There were no web pages, because the web had not been invented. Google would not be founded for another decade. Twitter would not be founded for another two decades. Mark Zuckerberg, who would grow up to start Facebook, was two years old. In talking about advances in computing, people often focus on the equipment. Certainly the advances in computing equipment since 1986 have been spectacular. Compared to the high-end supercomputers of 1986, today's mobile phones have more memory, more computing horsepower, and a better network connection not to mention a vastly lower price.

[113] 18 U.S.C. §2510-2522.

[114] 18 U.S.C. §2511.

[115] 18 U.S.C. §§2701-2712.

[116] 18 U.S.C. §2701.

[117] 18 U.S.C. §§3121-3126. A trap and trace device identifies the source of incoming calls, and a pen register indicates the numbers called from a particular phone.

[118] 18 U.S.C. §3121.

[119] J. Beckwith Burr, "The Electronic Communications Privacy Act of 1986: Principles for Reform," March 30, 2010, http://www.digitaldueprocess.org/files/DDP_Burr_Memo.pdf.

[120] "Cloud computing is an emerging form of computing that relies on Internet-based services and resources to provide computing services to customers, while freeing them from the burden and costs of maintaining the underlying infrastructure. Examples of cloud computing include web-based e-mail applications and common business applications that are accessed online through a browser, instead of through a local computer" (Government Accountability Office, *Information Security: Federal Guidance Needed to Address Control Issues with Implementing Cloud Computing*, GAO-10-513, May 2010, http://www.gao.gov/new.items/d10513.pdf).

[121] House Committee on the Judiciary, Subcommittee on the Constitution, Civil Rights, and Civil Liberties, *ECPA Reform and the Revolution in Cloud Computing* (statement of Michael Hintze, Associate General Counsel, Microsoft Corp.).

[122] Digital Due Process Coalition, "Our Principles", 2010, http://www.digitaldueprocess.org/index.cfm?objectid= 99629E40-2551-11DF-8E02000C296BA163.

[123] House Republican Cybersecurity Task Force, *Recommendations*, p. 14. For more information on ECPA, see CRS Report 98-326, *Privacy: An Overview of Federal Statutes Governing Wiretapping and Electronic Eavesdropping*, by Gina Stevens and Charles Doyle.

[124] Prepared by John Rollins, Specialist in Terrorism and National Security (jrollins@crs.loc.gov, 7-5529).

[125] CRS Report RL31787, *Information Operations, Cyberwarfare, and Cybersecurity: Capabilities and Related Policy Issues*, by Catherine A. Theohary.

[126] 50 U.S.C. §413b(e) defines a covert action as "an activity or activities of the United States Government to influence political, economic, or military conditions abroad, where it is intended that the role of the United States Government will not be apparent or acknowledged publicly, but does not include ... activities the primary purpose of which is to acquire intelligence ... [or] traditional military activities or routine support to such activities."

[127] For an explanation and analysis of issues relating to covert and clandestine activities see CRS Report RL33715, *Covert Action: Legislative Background and Possible Policy Questions*, by Richard F. Grimmett.

[128] Parts of the chapter have also been given other popular names: the Next Generation Internet Research Act of 1998 (P.L. 105-305), and the Department of Energy High-End Computing Revitalization Act of 2004.

[129] Prepared by Patricia Moloney Figliola, Specialist in Internet and Telecommunications Policy (pfigliola@crs.loc.gov, 7-2508).

[130] House Republican Cybersecurity Task Force, *Recommendations*, p. 14.

[131] Prepared by Catherine A. Theohary, Analyst in National Security Policy and Information Operations (ctheohary@crs.loc.gov, 7-0844). These provisions are codified to Chapter 5 of Title 47, the "Communications Act of 1934." Codification of the various provisions of this act is complex. See 47 U.S.C. §609 nt. for details.

[132] See CRS Report R41499, *The Communications Decency Act: Section 230(c)(1) and Online Intermediary Liability*, by Kathleen Ann Ruane and Julia Tamulis.

[133] Prepared by Wendy R. Ginsberg, Analyst in Government Organization and Management (wginsberg@crs.loc.gov, 7-3933), and Eric A. Fischer. The two divisions, originally known as the Federal Acquisition Reform Act and the Information Technology Management Reform Act, were renamed as the Clinger-Cohen Act by P.L. 104-208 and reclassified into 40 U.S.C. Subtitle III by P.L. 107-217 (see 40 U.S.C. §101 nt.).

[134] The Clinger-Cohen Act originally gave this promulgation authority to the Secretary of Commerce, while providing the President authority to disapprove or modify such standards, and gave the Secretary authority to waive the standards in specific cases to avoid adverse financial or mission-related impacts. The "Federal Information Security Management Act of 2002 (FISMA)", enacted as part of the Homeland Security Act, transferred that authority to OMB.

[135] See the discussion of FISMA, p. 42.

[136] They do have authority under FISMA to ensure compliance with that law's information security requirements (44 U.S.C. §3544). Some agency CIOs also have statutory authority in addition to that provided by Clinger-Cohen and FISMA. For example, the CIO of the

intelligence community has procurement approval authority for IT (50 U.S.C. §403-3g), and CIOs within DOD have budgetary review authority (10 U.S.C. §2223).

[137] Vivek Kundra, *25-Point Implementation Plan to Reform Federal Information Technology Management* (The White House, December 9, 2010), http://www.cio.gov/documents/25-Point-Implementation-Plan-to-ReformFederal%20IT.pdf.

[138] Jacob J. Lew, "Chief Information Officer Authorities," Memorandum for the Heads of Executive Departments and Agencies, M-11-29, August 8, 2011, pp. 1–2, http://www.whitehouse.gov/sites/default/files/omb/memoranda/2011/ m11-29.pdf.

[139] See CRS Report R40150, *A Federal Chief Technology Officer in the Obama Administration: Options and Issues for Consideration*, by John F. Sargent Jr.

[140] See, for example, CRS Report RL30914, *Federal Chief Information Officer (CIO): Opportunities and Challenges*, by Jeffrey W. Seifert.

[141] Prepared by Kristin M. Finklea, Coordinator, Analyst in Domestic Security (kfinklea@crs.loc.gov, 7-6259). See 18 U.S.C. §1001 nt. for classification details.

[142] The FTC now records consumer complaint data and reports it in the Identity Theft Data Clearinghouse (Federal Trade Commission, "Reference Desk," *Fighting Back Against Identity Theft*, December 22, 2010, http://www.ftc.gov/ bcp/edu/microsites/idtheft /reference-desk/index.html); identity theft complaint data are available for 2000 and forward.

[143] For classification details, see 6 U.S.C. §101 nt.

[144] In particular, the act transferred to DHS the Federal Computer Incident Response Center, which had resided in the General Services Administration (GSA). In 2006, P.L. 109-295, The Department of Homeland Security Appropriations Act, 2007, established the position of Assistant Secretary for Cybersecurity and Communications (6 U.S.C. §321) within DHS but did not specify duties or responsibilities.

[145] This set of provisions (Subtitle G of Title VIII, 6 U.S.C. §441-444) is called the SAFETY Act.

[146] This set of provisions (Subtitle I of Title VIII, 6 U.S.C. §481-486) is called the Homeland Security Information Sharing Act. Sec. 486 was added by P.L. 109-90 and provides some liability protections relating to actions involving information sharing and analysis centers.

[147] See also "Communications Act of 1934" above.

[148] This center would presumably replace the federal incident center currently required under 44 U.S.C. 3546. The revision of the Federal Information Security Management Act of 2002 (FISMA) in the *White House Proposal* does not include the latter center.

[149] FISMA was originally enacted as part of the Homeland Security Act of 2002, replacing provisions enacted by the Floyd D. Spence National Defense Authorization Act for Fiscal Year 2001 (P.L. 106-398, Title X, Subtitle G), enacted in 2000 but with a 2002 sunset. FISMA was reenacted in the same Congress by the E-government Act. Subchapter II is not in effect. The title 40 provision was originally enacted as part of the Clinger-Cohen Act (see p. 38), and the title 15 provisions are part of the NIST Act (see p. 23). See footnote 151 for more detail.

[150] For a more detailed description, see, for example, Government Accountability Office, *Information Security: Weaknesses Continue Amid New Federal Efforts to Implement Requirements*, GAO-12-137, October 2011, http://www.gao.gov/new.items/d12137.pdf.

[151] The standards-promulgation authority had been granted to the Secretary of Commerce under the Clinger-Cohen Act of 1996 (P.L. 104-106) but was transferred to the Director of OMB by the FISMA title in the HSA in 2002 (P.L. 107- 296, Sec. 1002, 40 U.S.C. 11331). The version of the main Chapter 35 currently in effect (Subchapter III) was enacted by the FISMA title in the E-Government Act of 2002 (P.L. 107-347, Title III), which suspended

Subchapter II, which had been revised by the HSA. That is not the case for 40 U.S.C. 11331, for which the P.L. 107-347 version would have retained the authority of the Secretary of Commerce to promulgate those standards as established in the Clinger-Cohen Act of 1996 (see p. 38), even though the E-Government Act was enacted after the HSA. Similarly, the revision to the NIST Act at 15 U.S.C. 278g-3 & 4 is that made by the HSA. The reason for this potentially confusing difference appears to be that (1) the effective date of HSA was later than that of the E-Government Act, and (2) HSA amended the existing subchapter II of 44 U.S.C. Chapter 35; the E-Government Act explicitly suspended that subchapter. In contrast, the revisions both laws made to the Paperwork Reduction Act, adding a subsection (c) to 44 U.S.C. §3505 (requiring inventories of federal information systems) were codified. However, there appear to remain some ambiguities in interpretation of the applicability of the two acts, which would presumably be resolved if FISMA is revised.

[152] See, for example, S.Rept. 111-368, and House Subcommittee on Government Management, Organization, and Procurement, *The State of Federal Information Security, Committee on Oversight and Government Reform* (Washington, DC: U.S. Government Printing Office, 2009), http://www.gpo.gov/fdsys/pkg/CHRG-111hhrg57125/pdf/ CHRG-111 hhrg 57125.pdf. OMB has recently attempted to address some of the operational issues administratively by delegating some responsibilities to DHS (Orszag and Schmidt, "Clarifying Cybersecurity Responsibilities and Activities of the Executive Office of the President and the Department of Homeland Security (DHS)"). Weaknesses in FISMA implementation have been cited repeatedly by GAO in reports required by the act (see, for example, Government Accountability Office, *Information Security: Weaknesses Continue Amid New Federal Efforts to Implement Requirements*).

[153] The bill included provisions from H.R. 4900, which was ordered reported by the House Oversight and Government Reform Committee.

[154] House Republican Cybersecurity Task Force, *Recommendations*, p. 13.

[155] The original act was amended by P.L. 109-144, the Terrorism Risk Extension Act of 1995, and P.L. 110-160, the Terrorism Risk Insurance Program Reauthorization Act of 2007. For classification details, see 15 U.S.C. 6701 nt.

[156] See, for example, Karen C. Yotis, "TRIA and the Perils of Terrorism Insurance," *Viewpoint*, Summer 2007, http://www.aaisonline.com/viewpoint/07sum6.html.

[157] 15 U.S.C. §§278g,h are part of the NIST Act (see p. 23).

[158] See, for example, NITRD, "About the NITRD Program: National Cyber Leap Year", July 22, 2009, http://www.nitrd.gov/leapyear/index.aspx.

[159] The White House, *Cyberspace Policy Review*.

[160] See, for example, H.Rept. 111-405, CSIS Commission on Cybersecurity for the 44th Presidency, *A Human Capital Crisis in Cybersecurity*, July 2010, http://csis.org/files/ publication/ 100720_Lewis_HumanCapital_WEB_BlkWhteVersion.pdf.

[161] It would include "any information about an individual maintained by an agency."

[162] Prepared by Kristin M. Finklea, Analyst in Domestic Security (kfinklea@crs.loc.gov, 7-6259). For classification details, see 18 U.S.C. §1028 nt.

[163] Examples of such federal crimes include theft of public property, theft by a bank officer or employee, theft from employee benefit plans, false statements regarding Social Security and Medicare benefits, several fraud and immigration offenses, and specified felony violations pertaining to terrorist acts.

[164] For more information on identity theft, see CRS Report R40599, *Identity Theft: Trends and Issues*, by Kristin M. Finklea.

[165] Federal Trade Commission, *Consumer Sentinel Network Data Book for January–December, 2010*, March 2010, http://www.ftc.gov/sentinel/reports/sentinel-annual-reports/sentinel-cy2010.pdf.

[166] Javelin Strategy & Research, *2011 Identity Fraud Survey Report: Consumer Version*, February 2011, p. 5 (available at https://www.javelinstrategy.com/brochure/207).

[167] The President's Identity Theft Task Force, *Combating Identity Theft: A Strategic Plan*, April 2007, http://www.identitytheft.gov/reports/StrategicPlan.pdf.

[168] This would involve revision of 18 U.S.C. §§1028 and 1028A.

[169] A predicate offense can be described as a crime that is a component of a more serious offense. For example, in the case of money laundering, the crime that produces the funds that are to be laundered is the predicate offense.

[170] This would involve revision of 18 U.S.C. §1028A.

[171] The President's Identity Theft Task Force, *Combating Identity Theft: A Strategic Plan*, pp. 91 – 92.

[172] 18 U.S.C. §1708.

[173] 18 U.S.C. §513.

[174] 26 U.S.C. §7201, 7206-7207.

[175] House Republican Cybersecurity Task Force, *Recommendations*, p. 14.

[176] Prepared by John Rollins, Specialist in Terrorism and National Security (jrollins@crs.loc.gov, 7-5529). Classification of this act is complex. For details, see 50 U.S.C. §401 nt.

[177] National Commission on Terrorist Attacks Upon the United States , *The 9/11 Commission Report*, July 22, 2004, http://www.9-11commission.gov/report/911Report.pdf.

[178] For more information on threats, responses, and issues associated with cyberterrorism, see CRS Report R41674, *Terrorist Use of the Internet: Information Operations in Cyberspace*, by Catherine A. Theohary and John Rollins.

In: Cybersecurity and Related Federal Laws ISBN 978-1-62257-743-9
Editors: J. R. Henderson and D. I. Cox

Chapter 2

CYBERSECURITY: SELECTED LEGAL ISSUES*

Edward C. Liu, Gina Stevens, Kathleen Ann Ruane, Alissa M. Dolan and Richard M. Thompson II

SUMMARY

The federal government's role in protecting U.S. citizens and critical infrastructure from cyber attacks has been the subject of recent congressional interest. Critical infrastructure commonly refers to those entities that are so vital that their incapacitation or destruction would have a debilitating impact on national security, economic security, or the public health and safety. This report discusses selected legal issues that frequently arise in the context of recent legislation to address vulnerabilities of critical infrastructure to cyber threats, efforts to protect government networks from cyber threats, and proposals to facilitate and encourage sharing of cyber threat information among private sector and government entities. This report also discusses the degree to which federal law may preempt state law.

It has been argued that, in order to ensure the continuity of critical infrastructure and the larger economy, a regulatory framework for selected critical infrastructure should be created to require a minimum level of security from cyber threats. On the other hand, others have argued that such regulatory schemes would not improve cybersecurity

* This is an edited, reformatted and augmented version of a Congressional Research Service publication, CRS Report for Congress R42409, from www.crs.gov, prepared for Members and Committees of Congress, dated April 20, 2012.

while increasing the costs to businesses, expose businesses to additional liability if they fail to meet the imposed cybersecurity standards, and increase the risk that proprietary or confidential business information may be inappropriately disclosed.

In order to protect federal information networks, the Department of Homeland Security (DHS), in conjunction with the National Security Agency (NSA), uses a network intrusion system that monitors all federal agency networks for potential attacks. Known as EINSTEIN, this system raises significant privacy implications—a concern acknowledged by DHS, interest groups, academia, and the general public. DHS has developed a set of procedures to address these concerns such as minimization of information collection, training and accountability requirements, and retention rules. Notwithstanding these steps, there are concerns that the program may implicate privacy interests protected under the Fourth Amendment.

Although many have argued that there is a need for federal and state governments, and owners and operators of the nation's critical infrastructures, to share information on cyber vulnerabilities and threats, obstacles to information sharing may exist in current laws protecting electronic communications or in antitrust law. Private entities that share information may also be concerned that sharing or receiving such information may lead to increased civil liability, or that shared information may contain proprietary or confidential business information that may be used by competitors or government regulators for unauthorized purposes.

Several bills in the 112th Congress would seek to improve the nation's cybersecurity, and may raise some or all of the legal issues mentioned above. For example, H.R. 3523 (Rogers (Mich.)-Ruppersberger) addresses information sharing between the intelligence community and the private sector. H.R. 3674 (Lungren) includes provisions regarding the protection of critical infrastructure, as well as information sharing. H.R. 4257 (Issa-Cummings) would require all federal agencies to continuously monitor their computer networks for malicious activity and would impose additional cybersecurity requirements on all federal agencies. S. 2102 (Feinstein) seeks to facilitate information sharing. S. 2105 (Lieberman) includes the information sharing provisions of S. 2102, as well as provisions relating to the protection of critical infrastructure and federal government networks. S. 2151 (McCain) and H.R. 4263 (Bono-Mack) also addresses information sharing among the private sector and between the private sector and the government. Many of these bills also include provisions specifically addressing the preemption of state laws.

INTRODUCTION

For many, the Internet has become inextricably intertwined with daily life. Many rely on it to perform their jobs, pay their bills, send messages to loved ones, track their medical care, and voice political opinions, among a host of other activities. Likewise, government and business use the Internet to maintain defense systems, protect power plants and water supplies, and keep other types of critical infrastructure running.[1] Consequently, the federal government's role in protecting U.S. citizens and critical infrastructure from cyber attacks has been the subject of recent congressional interest.[2]

This report discusses selected legal issues that frequently arise in the context of legislation to address vulnerabilities of private critical infrastructure to cyber threats, efforts to protect government networks from cyber threats, and proposals to facilitate and encourage sharing of cyber threat information amongst private sector and government entities. This report also provides an overview of the ways in which federal laws of these types may preempt or affect the applicability of state law.

LEGAL ISSUES RELATED TO PROTECTING CRITICAL INFRASTRUCTURE

Although no federal statute currently imposes a generally applicable obligation on businesses in the private sector to take measures to protect themselves from cyber vulnerabilities, Congress has chosen to impose regulatory standards regarding the security, including the cybersecurity, of specific sectors or types of private entities.[3] For example,[4] chemical facilities are subject to chemical facility anti-terrorism standards (CFATS) promulgated by the Department of Homeland Security (DHS), which include provisions requiring chemical facilities to take measures to protect against cyber threats.[5] Electrical utilities are required to comply with reliability standards, including standards to protect against cyber incidents, set by the North American Electrical Reliability Corporation (NERC).[6] Similarly, the Maritime Transportation Security Act (MTSA) gives the Coast Guard the authority to regulate the security of maritime facilities and vessels, including requiring security plans that contain provisions for the security of communications systems used in those facilities.[7]

Proposals that focus on the increased cybersecurity of certain sectors of the economy are frequently justified on the grounds that those private entities, including energy, transportation, or communication providers, comprise the nation's critical infrastructure. If the incapacity or destruction of such systems or assets would have a debilitating impact on national security, economic security, or public health and safety, it would be in the national interest to ensure that such critical infrastructure was adequately protected. Consequently, it has been argued that a regulatory framework governing selected critical infrastructure entities is needed to ensure that these private entities take measures adequate to maintain a minimum level of security from cyber threats, in order to protect the rest of the economy.[8]

On the other hand, others have argued that such regulatory schemes would not improve cybersecurity and would also increase the costs of doing business for these sectors of the economy.[9] There are also concerns that businesses would face additional exposure to civil liability from private suits if they failed to meet the imposed standards. As many of these regulatory schemes provide regulatory agencies with access to information held by the regulated entities, concerns have also been raised about the inappropriate disclosure of proprietary or confidential business information.

The concerns raised by these issues have shaped the existing legal schemes regulating the security of specific categories of critical infrastructure, and have also informed recent legislative proposals to address widely reported weaknesses in the security of critical infrastructure from cyber threats. A brief overview of each of these issues is provided in the next sections of this report. The report will then examine how recent cybersecurity legislation would specifically address some or all of these issues.

Deference to Agency Decisions

Several of the bills that would establish a regulatory scheme for the cybersecurity of critical infrastructure provide the agencies charged with administering the program with the discretion to identify those private entities that would fall within the scope of a particular bill and that will, therefore, be subject to the requirements that would be imposed under the bill. Being subject to the regulations may have significant cost, liability, or other implications for the private entity that has been designated as covered critical infrastructure; such entities may seek to challenge their designation as covered critical infrastructure through redress mechanisms created in the statute or through judicial review of agency action under the Administrative Procedure Act (APA).[10] Entities may also seek judicial review of agency actions in the context of enforcement actions taken against them under the various regulatory schemes.

Depending upon the legislative language delegating regulatory authority to the agency, a court will evaluate an agency's decision under varying standards of review. In the context of regulating the security of critical infrastructure, a more deferential standard of review of agency determinations typically means that regulated private entities would have less recourse in the event that they wanted to challenge an agency's determination. On the other hand, a less deferential standard of review may extend the time to implement particular security standards if the agency encounters delays caused by litigation. Examples of the different types of judicial review that may be involved are discussed below.

Availability of Judicial Review[11]

As a general matter, there is a "'strong presumption that Congress intends judicial review' of administrative action."[12] This presumption is embodied in the Administrative Procedure Act (APA), which provides that "final agency action for which there is no other adequate remedy in a court [is] subject to judicial review."[13] The APA provides two exceptions to the presumption of availability of judicial review of agency action: (1) "to the extent that ... statutes preclude judicial review" and (2) "where agency action is committed to agency discretion by law."[14] However, judicial review of an unreviewable determination may occur if there is a constitutional issue.[15]

Under the APA, judicial review of agency actions may be unavailable if such review is specifically precluded by statute.[16] This exemption requires the existence of an explicit statutory provision prohibiting judicial review of agency action. Additionally, even where judicial review has not been explicitly barred, the APA precludes judicial review where the decision has been committed to agency discretion by law.[17] This second exemption has been interpreted by the Supreme Court to be a very narrow exception, and applies only in situations where the statute provides no law for a reviewing court to apply.[18] For example, in *Webster v. Doe*,[19] the Supreme Court held that firing decisions made by the Director of Central Intelligence were unreviewable because the National Security Act provided that the Director "may, in his discretion, terminate the employment of any officer or employee of the [Central Intelligence Agency] whenever he shall deem such termination necessary or advisable in the interests of the United States."20 The Court held that such a statute "exuded deference" and noted:

> Short of permitting cross-examination of the Director concerning his views of the Nation's security and whether the discharged employee was inimical to those interests, we see no basis on which a reviewing court could properly assess an Agency termination decision.[21]

Since the statute contained no standards a court could apply to evaluate the Director's decision, the Court determined that these decisions had been committed to agency discretion by law, and were consequently unreviewable.

Questions of Fact

Where a statute does provide judicially administrable standards, agency determinations of factual questions are typically reviewed under the "substantial evidence" or "abuse of discretion standards."[22] In the administrative context, substantial evidence review and abuse of discretion review occur in factually distinct circumstances. Substantial evidence is required when an agency engages in either formal rulemaking or an adjudicatory hearing.[23] In contrast, abuse of discretion applies in cases of informal rulemaking and decisions.[24]

Some courts appear to consider substantial evidence a more demanding standard than abuse of discretion, but the consistent theme of both standards is that the court is not free to substitute its judgment in place of the agency's.[25] In terms of analysis, the substantial evidence and abuse of discretion standards are both less stringent than *de novo* review, which would allow a court to look at the evidence anew and come to its own conclusions. Nevertheless, the Supreme Court has described these standards as requiring "more than a mere scintilla" of support and comparable to the standard a trial judge must meet to sustain a jury's verdict.[26] In the federal courts, a jury verdict will not be disturbed if "reasonable and fair-minded persons in exercise of impartial judgment" might have come to the same conclusion as the jury.[27]

Interpretations of Law

Agencies may also exercise discretion in interpreting the terms used in a statute. In the context of the proposals to regulate the cybersecurity of critical infrastructure, which are discussed in more detail below, there are a number of provisions that may require the Secretary of Homeland Security (the Secretary) to use her discretion to interpret the language of the bills. For example, the various definitions for covered critical infrastructure used by the bills may require a finding that the disruption of a function, system, or asset would lead to harms that were "significant," "extraordinary," or "prolonged." These terms may be susceptible to more than one specific construction, and the different interpretations may have material consequences for those subject to the regulatory scheme. A narrow definition may mean that fewer entities would be subject to regulation, while a broader definition may encompass a more expansive cross-section of businesses.

The validity of an agency's construction of a statute would likely be evaluated using the two-prong test described by the Supreme Court in *Chevron v. Natural Resources Defense Council.*[28] First, if the text and legislative history of the statute demonstrate that Congress has spoken directly on the issue, then that statutory language or history must control. However, under the second prong, if the statute is ambiguous because "Congress has not directly addressed the precise question at issue," the agency's interpretation will stand so long as it is a reasonable one.[29]

Therefore, under *Chevron*, whether a particular statutory provision is ambiguous or not can change the degree of deference afforded an agency. Where no ambiguity exists, the reviewing court's focus is on the intent of Congress, and it may interpret the law *de novo* without any deference toward the agency's interpretation. On the other hand, if the statute is ambiguous, either because the language used is susceptible to more than one meaning or because the law contains internal inconsistencies, the reviewing court is not permitted to supplant its own interpretive preferences for that of the agency, unless the agency's interpretation is unreasonable. Under this deferential standard of review, the discretion available to an agency is inversely proportional to the degree of specificity provided in a particular statute. In the context of the bills discussed by this report, the less specific a particular bill is regarding the Secretary's regulatory authority, the more flexibility would be available to her to exercise during implementation.

Liability Concerns

The creation of a regulatory scheme applicable to critical infrastructure may raise issues regarding the effects that the new regulatory scheme would have on the potential civil or criminal liability of the covered entities. Regulators may be given the authority to impose civil or criminal penalties for noncompliance, or may seek to promote compliance by offering financial incentives.[30]

In addition to the forms of liability imposed by regulatory authorities, questions may arise regarding the potential ways in which the regulatory scheme may expose covered entities to additional private civil liability. In this context, a federal regulatory scheme could be viewed as creating a standard of care that might be used to establish tort liability under state law. Entities that fall below that standard of care face the possibility of liability in the event of a security breach, separate and apart from any penalties that might be imposed by government regulators. The most likely form that such a civil action would take is in a tort suit alleging that the private entity had acted negligently; that is, the entity had failed to exercise reasonable care in the face of a foreseeable risk. Under current state law, entities found negligent may be liable for harm that results from their negligence.[31]

The existence of a federal regulatory scheme that imposes compliance standards may affect suits alleging negligence in two ways. First, the entities that are subject to the compliance standards may be found negligent *per se* if they fail to satisfy those standards.[32] Negligence *per se* is a theory of negligence in which the fact that an entity's conduct has violated some applicable statute is *prima facie* evidence that the entity has acted negligently.[33] Unless the defendant could rebut that presumption, the defendant would likely be found to be *per se* negligent, and consequently liable for any harm that results from that negligence.[34] In the context of cyber threats to critical infrastructure, this might mean that a regulated entity that fails to adequately secure its information infrastructure as required under a federal regulatory scheme would be liable for a cyber incident that causes harm to customers or other third parties.

Second, entities that are not subject to regulation under a federal scheme may not be subject to negligence *per se*. However, the performance standards or other requirements imposed under that scheme may still affect their liability for negligence if such requirements establish an applicable standard of care that the nonregulated entity would be judged against in a private civil suit.[35]

Freedom of Information[36]

Access to the confidential business information of owners and operators of the nation's critical infrastructure and of private sector entities continues to be an important component of efforts to protect against cybersecurity threats.

However, some critical infrastructure owners and operators and private sector entities may be hesitant to share cybersecurity-related information with the government because of the possible disclosure of this information to the public under the Freedom of Information Act (FOIA)[37] and state open records laws.[38] In addition, concerns also exist that sharing of cybersecurity information may facilitate access to proprietary and confidential business information by competitors. Furthermore, some have expressed concerns that the government may use information obtained for cybersecurity purposes for noncybersecurity purposes, such as regulatory actions. Concerns also exist that reliance on FOIA's exemptions to shield shared cybersecurity threat information is misplaced because court interpretations of the scope of FOIA's exemptions can change.[39] Proponents of open records and government transparency argue that new exemptions from FOIA jeopardize the public's ability to obtain information about government and industry practices, cast a shroud of secrecy over government's functions, and are unnecessary because FOIA's exemptions adequately protect private information from disclosure.[40] Some observers believe that it is not certain that some cybersecurity threat information, such as routing information or website access logs, would fit within FOIA's exemptions.

The Freedom of Information Act of 1974 (FOIA) regulates the disclosure of federal agency records.[41] FOIA requires that certain types of records be published in the *Federal Register*;[42] that certain types of records be made available for public inspection and copying;[43] and that all other records be subject to request in writing. All records not available via publication or inspection, not exempt from disclosure, or excluded from coverage are subject to disclosure.[44] FOIA has nine exemptions from disclosure which permit, rather than require, the withholding of the requested information.[45]

Subsection (b)(3) of FOIA, commonly referred to as exemption 3, permits agencies to withhold information under FOIA that is specifically prohibited from disclosure by other federal statutes.[46] For a nondisclosure provision in a separate federal statute to qualify for exemption 3 status, the nondisclosure provision must meet the following criteria: either the statute must require that matters be withheld from the public in such a manner as to leave no discretion on the issue; or the statute must establish particular criteria for withholding or refer to particular types of matters to be withheld; and it must specifically cite FOIA exemption 3.[47] If the statute meets the criteria of exemption 3 and the information to be withheld falls within the scope and coverage of FOIA, the information is exempt from disclosure under exemption 3.[48] Statutes that meet these criteria are referred to as "FOIA exemption 3 statutes."[49]

To encourage private and public sector entities and persons to voluntarily share their critical infrastructure information with the Department of Homeland Security (DHS), the Critical Infrastructure Information Act of 2002 (CIIA) includes several measures to ensure against disclosure of protected critical infrastructure information by DHS. According to the Department of Justice, the agency responsible for administering FOIA, the CIIA will operate as an exemption 3 statute under FOIA for critical infrastructure information that is obtained by the Department of Homeland Security.[50] Relevant to this discussion, the CIIA provides protections against the disclosure of information that is voluntarily submitted by a critical infrastructure entity to DHS. If the information submitted satisfies the requirements of the CIIA, the information is designated as critical infrastructure information (CII), and for purposes of FOIA, the CIIA expressly prohibits the disclosure of critical infrastructure information. Critical infrastructure information "means information not customarily in the public domain and related to the security of critical infrastructure or protected systems...."[51] Therefore, the classification of information as CII would protect that information from disclosure under FOIA, state and local disclosure laws, and use in civil litigation. In addition, protected critical infrastructure information cannot be used for regulatory purposes.[52] Federal, state, and local government officials and contractors approved by DHS can access the information for critical infrastructure protection or criminal law enforcement purposes.

With respect to concerns about litigation, CIIA limits the use of CII in civil litigation and provides that sharing CII with the agency does not count as the "waiver of any applicable privilege or protection provided under law," such as trade secret protection or the attorney-client privilege.[53] CIIA authorizes the use or disclosure of such information by officers and employees in furtherance of the investigation or the prosecution of a criminal act, or for disclosure to Congress or the Government Accountability Office.

Another Exemption 3 statute under FOIA for critical infrastructure information was recently enacted in the National Defense Authorization Act for Fiscal Year 2012. Section 1091 authorizes the Secretary of Department of Defense (DOD), or his designee, to exempt DOD critical infrastructure security information from disclosure pursuant to section 552(b)(3) of title 5 [FOIA Exemption 3] upon a written determination that the information is DOD critical infrastructure security information, and the public interest consideration in the disclosure of such information does not outweigh preventing the disclosure of such information.54 Department of Defense critical infrastructure security information means sensitive but unclassified information that, if disclosed, would reveal vulnerabilities of DOD critical infrastructure that could result in the disruption, degradation, or destruction of Department of Defense (DOD) operations, property, or facilities.

Ex Parte Communications

Providing information to a regulatory agency may also be subject to further disclosure if the communication would implicate agency rules or judicial doctrine regarding ex parte communications. Under the APA, formal agency adjudications are to be decided solely on the basis of record evidence. The APA provides that "[t]he transcript of testimony and exhibits, together with all papers and requests filed in the proceeding, constitutes the exclusive record for decision."[55] The reason for this "exclusiveness of record" principle is to provide fairness to the parties in order to ensure meaningful participation. Challenges to the "exclusiveness of record" occur when there are ex parte contacts—communications from an interested party to a decision-making official that take place outside the hearing and off the record.[56] The APA prohibits any "interested person outside the agency" from making, or knowingly causing, "any ex parte communication relevant to the merits of the proceeding" to any decision making official.[57] Similar restraints are imposed on the agency decision makers.[58] Additionally, ex parte communications received in violation of these rules are generally required to be disclosed to all other interested parties and made part of the public record for the proceeding.[59] The CIIA provides that CII will not be subject to agency rules or judicial doctrine regarding ex parte communications. However, if an entity is involved in a proceeding where ex parte communications are prohibited, there may be concerns that providing cybersecurity information that would not qualify as CII might implicate the rules against ex parte communications, and could be subject to disclosure on the public record or to other interested parties.

Legislation in the 112th Congress

This section provides a brief description of proposed cybersecurity legislation in the 112th Congress that includes regulatory provisions regarding the security of critical infrastructure with particular emphasis placed on the provisions that implicate the legal issues discussed above.

H.R. 3674, Precise Act of 2011

H.R. 3674, the Promoting and Enhancing Cybersecurity and Information Sharing Effectiveness Act of 2011 (PRECISE Act), was introduced on December 15, 2011, by Representative Lungren. On February 1, 2012, the Subcommittee on Cybersecurity, Infrastructure Protection, and Security Technologies of the House Homeland Security Committee held a mark up of the bill, and it was forwarded to the full committee by voice vote. On April 18, 2012, full committee consideration and mark up were held on a substitute bill that changed the critical infrastructure provisions. The measure was reported favorably to the full House (as amended) 16-13.

As introduced, Section 2 of the PRECISE Act would have authorized the Secretary of Homeland Security to identify and evaluate cybersecurity risks to critical infrastructure and to review and develop a collection of existing internationally recognized consensus-developed risk-based performance standards.[60] It would have given the Secretary the authority to designate particular facilities or functions of critical infrastructure companies as covered critical infrastructure.[61] Such entities would have been subject to more stringent regulation than noncovered critical infrastructure. Additionally, the PRECISE Act would have explicitly provided for judicial review of the designation of facilities or functions as covered critical infrastructure.[62]

On April 18, the full committee held a mark up and reported out a substitute bill. This version removed the proposed Section 227, which would have provided the Secretary its authority to designate particular facilities as critical infrastructure and develop the standards these entities would have been required to implement. Under the approved PRECISE Act, DHS may "solely upon the request of critical infrastructure owners and operators, assist such critical infrastructure owners and operators in protecting their critical infrastructure information systems."

S. 2105, Cybersecurity Act of 2012

S. 2105, the Cybersecurity Act of 2012, was introduced on February 14, 2012, by Senator Lieberman. Title I of the bill would create a regulatory scheme for the protection of selected systems and assets of critical infrastructure from cybersecurity threats. Specifically, the bill would authorize the Secretary of Homeland Security to identify cybersecurity risks to critical infrastructure,[63] designate certain assets or systems as "covered critical infrastructure,"[64] and identify performance standards that covered critical infrastructure would have to meet in order to guard against the identified cybersecurity risks.[65] If the Secretary determines that an existing regulatory scheme would adequately protect covered critical infrastructure from cyber threats, then no new performance standards would be imposed with respect to that covered critical infrastructure.[66]

The potential applicability of this new regulatory regime to an entity would depend upon whether its systems or assets had been designated by the Secretary as covered critical infrastructure.[67] The bill would define covered critical infrastructure as systems or assets that, if damaged or accessed without authorization, could reasonably lead to the interruption of life sustaining services sufficient to cause a mass casualty event with an extraordinary number of fatalities or mass evacuations with a prolonged absence, catastrophic economic damage to the United States, or severe degradation of national security.[68] Catastrophic economic damage is defined to include the failure or substantial disruption of a U.S. financial market, transportation system, or other systemic, long-term damage to the U.S. economy.[69] Commercial information technology products[70] are statutorily precluded from being designated as covered critical infrastructure, as are systems or assets based solely on activities that are protected by First Amendment rights.[71]

Because of a perceived increased regulatory burden that might accompany a designation of a system or asset as covered critical infrastructure, some entities may wish to dispute such a designation. The Cybersecurity Act of 2012 would explicitly provide for judicial review of decisions to designate systems or assets as covered critical infrastructure.[72] Such review would likely involve both questions of fact as well as interpretations of the bill's language. Examples of factual questions that might be raised include whether the disruption of an asset could lead to a mass casualty event or degradation of national security. The Cybersecurity Act of 2012 does not specify a particular standard of review that courts should use when reviewing these questions. But, under the APA, a court is likely to apply a "substantial evidence" or "abuse of discretion" standard to these types of factual questions.

Questions of law might also arise in the context of a designation as covered critical infrastructure under the Cybersecurity Act of 2012. For example, the bill does not provide specific definitions for terms such as an "extraordinary number of fatalities" or "prolonged absence." The bill would also prohibit the Secretary from designating a commercial information technology product, or any services provided in support of a commercial information technology product, as covered critical infrastructure.[73] Questions of interpretation may arise with respect to this exemption. For example, the Secretary may wish to designate a larger system, which happens to contain a commercial information technology product, as covered critical infrastructure.

However, the affected entity may argue that such a designation would violate the bill's prohibition on designating commercial information technology products as covered critical infrastructure. If the Secretary were to interpret these provisions as permitting that designation, perhaps arguing that there is a distinction between designating a commercial information technology product as critical infrastructure and designating a system that is partially comprised of a commercial information technology product as covered critical infrastructure, it is likely that a reviewing court would evaluate this interpretation under the *Chevron* analysis described above. Specifically, a reviewing court would first ask whether the statute clearly answered the question, and, if the statute did not, would uphold the Secretary's interpretation to the extent that it is a reasonable one.

In order to enforce its provisions, the Cybersecurity Act of 2012 explicitly authorizes DHS to develop civil monetary penalties to be levied against covered critical infrastructure that was found to be noncompliant with the applicable performance standards.[74] The bill would allow owners or operators of covered critical infrastructure to self-certify annually that they are compliant, or submit to a third-party assessment of compliance.[75] However, audits and inspections by DHS would also be authorized if there were a reasonable suspicion of noncompliance.[76]

With respect to private civil liability, the Cybersecurity Act of 2012 provides some immunity for covered critical infrastructure that experience cybersecurity incidents related to identified risks.[77] The owner or operator of the covered critical infrastructure would be eligible to receive immunity from punitive damages in a private civil suit, but such immunity would be available if the entity had also met applicable performance requirements under the bill, had received a successful assessment, and was also in substantial compliance at the time of the incident.[78]

The Cybersecurity Act of 2012 would authorize the Secretary of Homeland Security to collect information from covered critical infrastructure in order to conduct risk assessments and to evaluate compliance with applicable performance standards.[79] The bill provides that any information collected under its authority would be considered critical infrastructure information (CII) under the Critical Infrastructure Information Act of 2002 (CIIA).[80] While information must normally be submitted voluntarily in order to be considered CIIA, the Cybersecurity Act of 2012 removes this requirement with respect to information that would be collected pursuant to the bill.[81] Information would not be considered CII if it were submitted to conceal violations of law, inefficiency, or administrative error; prevent embarrassment to a person, organization, or agency; or interfere with competition in the private sector.[82]

In addition to the authorities established under Title I of the Cybersecurity Act of 2012, Title III of the bill would amend the Homeland Security Act of 2002 to create a National Center for Cybersecurity and Communications (NCCC or Center).[83] The NCCC is charged with managing "Federal efforts to secure, protect, and ensure the resiliency of the Federal information infrastructure, national information infrastructure, and national security and emergency preparedness communications infrastructure...."[84] The Director of the NCCC will be appointed by the President and will report directly to the Secretary of Homeland Security (the Secretary).[85] Additionally, the NCCC will have one deputy director from the intelligence community[86] who is chosen by the Director of National Intelligence and reports directly to the Secretary.[87]

There are several places in which Title I provisions detailing DHS's authority to regulate critical infrastructure security overlap with the Center's responsibilities as outlined in Title III. For example, both the Center and the Secretary of Homeland Security are instructed to conduct cyber risk assessments of critical infrastructure,[88] and inform critical infrastructure owners about security conditions. The Center must provide classified and unclassified security information to national information infrastructure owners, which could include entities designated as critical infrastructure by DHS under Title I.[89] DHS is required to provide information to critical infrastructure owners about cybersecurity threats, however, provision of classified information is not directly addressed.[90] Finally, both entities play a role in responding to cybersecurity emergencies. The Center must develop and coordinate a "national incident response plan that details the roles of Federal agencies, State and local governments, and the private sector...."[91] In similar language, DHS must "improve the capabilities and procedures of government and private section entities to respond to a major cyber incident" and "clarify specific roles, responsibilities, and authorities of the government" when responding.[92] Since these Title I and III authorities overlap but are not wholly duplicative, it may not be clear if or how the exercise of these authorities would coincide.

Title I also gives the President the authority to exempt organizations that have been designated as covered critical infrastructure by DHS from the requirements imposed in Title I, if they are sufficiently regulated by a sector-specific agency.[93] While these exempted entities are clearly free from Title I requirements, it appears that they are still subject to the Title III provisions that apply to covered critical infrastructure.[94] If they are not exempted, these information systems will be excepted from DHS requirements under Title I, but will still have to comply with the Title III affirmative obligation for covered critical infrastructure operators to share information with the Center about cyber incidents.[95]

Legal Issues Related to the Protection of Federal Networks

Prompted by a perceived threat to governmental information technology (IT) systems, DHS, in conjunction with the National Security Agency (NSA), has incrementally ramped up monitoring of federal government networks over the past decade to identify and prevent cyber attacks. A focal point of these efforts is EINSTEIN, a network intrusion system that monitors all federal agency networks for potential attacks. As part of this monitoring, all communications by federal executive agency employees made on federal networks, and incidentally, all communications they have with private citizens, are monitored for malicious activity. This monitoring may trigger Fourth Amendment guarantees to the right to be free from unreasonable searches and excessive government intrusion. Additionally, Congress has enacted statutory rules that place a higher restriction than the Constitution on government access to electronic communications.[96]

This section surveys EINSTEIN's background and discusses the Fourth Amendment concerns it raises for both federal employees and private citizen's communicating with them. It will then discuss alternative privacy and civil liberties protections that may be instituted to complement Fourth Amendment protections. Finally, this section discusses recent legislative efforts in the 112th Congress to improve the federal government's cybersecurity networks.

EINSTEIN Overview

Before EINSTEIN was introduced, federal agencies reported cyber threats to DHS manually and on an ad hoc basis.[97] It was usually done after the agency systems were affected by the attack. To remedy this, DHS, in collaboration with NSA, created EINSTEIN—a system to detect and report network intrusions. EINSTEIN's mandate derived from a combination of statutes, presidential directives, and agency memoranda.

The first mandates for EINSTEIN came in 2002 with the Homeland Security Act and Homeland Security Presidential Directive 7.[98] In 2007, the Office of Management and Budget required all federal executive agencies to develop a comprehensive plan of action to defend against cyber threats.[99] Coinciding with these statutory and administrative directives, DHS and NSA launched EINSTEIN in three phases, each increasingly more sophisticated than the last.

DHS rolled out EINSTEIN 1 in 2004 to automate the process by which federal agencies reported cyber threats to the United States Computer Emergency Readiness Team (US-CERT), the operational arm of DHS's cybersecurity division.[100]

Under EINSTEIN 1, federal agencies voluntarily sent "flow records" of Internet network activity to DHS so it could monitor the Internet traffic across the federal gov. domain. These flow records included basic routing information such as the IP addresses of the connecting computer and the federal computer connected to.[101] US-CERT used this information to detect and mitigate malicious activity that threatened federal networks. This information was shared with both public and private actors on the DHS website.[102]

In an effort to upgrade EINSTEIN's capabilities, DHS launched EINSTEIN 2, which is capable of alerting US-CERT of malicious network intrusions in near-real time.[103] Sensors installed at all federal agency Internet access points make a copy of all network activity coming to and from federal networks, including addressing information and the content of the communication.[104] These data are later scanned for the presence of "signatures," patterns that correspond to a known threat, such as denial of service attacks, network backdoors, malware, worms, Trojan horses, and routing anomalies.[105]

The system triggers an alert when it senses malicious activity. All the data corresponding with the trigger, including the content of the communication, are saved.[106] Personnel at US-CERT then analyze the stored messages and act accordingly.

In 2010, DHS began testing EINSTEIN 3 on one federal agency.[107] In addition to *detecting* cyber threats, this newest iteration also is designed to *block* and *respond* to these threats before any harm is done.[108] US-CERT is also testing the ability of EINSTEIN 3 to provide real-time information sharing with other federal agencies and the NSA.[109]

EINSTEIN and the Fourth Amendment

There is no doubt that EINSTEIN's monitoring of all communications coming to and from federal agency computers poses significant privacy implications—a concern acknowledged by DHS, interest groups, academia, and the general public.[110] This program affects not only federal employees, but also any private citizen who communicates with them. DHS has developed a set of procedures to address these concerns, such as minimization of information collection, training and accountability requirements, and retention rules. Notwithstanding these steps, growth of this Internet monitoring program may trigger privacy interests protected under the Fourth Amendment.

The Fourth Amendment provides in relevant part: "The right of the people to be secure in their persons, houses, papers, and effects, against unreasonable searches and seizures, shall not be violated...."[111] The principal purpose of the Fourth Amendment is to protect the privacy of individuals against invasion from government officials.[112] Not all government acts, however, trigger Fourth Amendment protections. For the Fourth Amendment to apply, a court must first inquire whether the governmental act constitutes a *search* or *seizure* in the constitutional sense.[113] To determine if a *search* has occurred, a court will ask whether the individual had an actual expectation of privacy that society would deem reasonable.[114] If yes, the court will then ask if the search was reasonable—the core Fourth Amendment requirement.[115] Except in well-defined instances, a search is not reasonable unless the government obtains a warrant based upon probable cause.[116] There are, however, exceptions to this rule such as special needs and consent that will be explored below.

There seems to be a consensus in federal courts that Internet users are not entitled to privacy in the non-content, routing information of their Internet communications.[117] In *United States v. Forrester*, the government obtained court permission to install a device similar to a pen register to record the to/from addresses of the defendant's emails, the IP addresses of the sites he visited, and the total volume of data sent to and from his account.[118] The Ninth Circuit Court of Appeals held that these surveillance techniques were indistinguishable from the pen register upheld by the Supreme Court in *Smith v. Maryland*.[119] Internet users should be aware, the panel reasoned, that this routing information is provided to the Internet service provider for the purpose of directing the information.[120]

On the other hand, the cases generally demonstrate that an individual has a legitimate expectation of privacy in the *content* of a communication. In

United States v. Warshak, the Ninth Circuit ruled that a "subscriber enjoys a reasonable expectation of privacy in the contents of emails that are stored with, or sent or received through, a commercial ISP."[121] In an earlier case, the Second Circuit opined that Internet users have an expectation of privacy in the content of the e-mail while in transmission.[122] Although the Supreme Court declined to resolve this issue in *City of Ontario v. Quon*, deciding the case on other grounds, it opined in dicta that "cell phones and text message communications are so pervasive that some persons may consider them to be an essential means or necessary instruments for self-expression, even self-identification. That might strengthen the case for an expectation of privacy."[123]

This content/non-content distinction is as old as Fourth Amendment case law.[124] In the late nineteenth century, the Court explained in *Ex parte Jackson* that the outside of a mailed letter—its "outward form and weight"—was not entitled constitutional protection.[125] However, the government must obtain a warrant before examining the contents of a letter or sealed package.[126] The Court protected the inside contents of the letter, but held that the outside, non-content material was not entitled to (in modern parlance) a reasonable expectation of privacy. This same rule was carried over to the telephone context.[127]

In *Katz v. United States*, the Court held that the contents of Katz's conversation—the actual words spoken—were protected under the Fourth Amendment.[128] A decade later the Court completed the other side of the doctrine in *Smith v. Maryland*, and held that a person has no expectation of privacy in the non-content, routing information of the telephone call—the numbers dialed.[129]

EINSTEIN 2 not only collects the routing, non-content portions of communications, such as email header information, but also scans and collects the content of the communications, such as the body of e-mails.[130] Based on the reasoning of the Internet content cases, individuals most likely have a reasonable expectation of privacy in those electronic communications.[131] The EINSTEIN program requires a Fourth Amendment inquiry into two discrete classes of individuals: (1) federal agency employees who access federal networks while at work; and (2) private persons who either contact a federal agency directly or who communicate via the Internet with a federal employee.[132]

The Fourth Amendment rights of the former primarily rest on cases dealing with privacy in the workplace and consent, while the latter requires a broader look at privacy and electronic communications.

Monitoring Communications from Federal Employees

As work and personal lives can become enmeshed, many employees are accessing not only work e-mail while on the clock, but also personal e-mails. EINSTEIN monitors not only federal executive agency employees' work e-mails or other official Internet activity, but also any information accessed on a federal agency computer including personal e-mails accessed from sites such as Gmail or Hotmail, or other Internet communications such as Facebook and Twitter. This poses several Fourth Amendment issues.

In *City of Ontario v. Quon*, the Supreme Court upheld under the Fourth Amendment the city's search of text messages sent on a city-issued pager by a police officer employed by that city.[133] Before issuing the pagers, the city had announced a usage policy that informed the officers that the city reserved the right to monitor the use of the pager including e-mail and Internet use, with or without notice to the employee.[134] The Court assumed without deciding that the employee had a reasonable expectation of privacy in the sent text messages, that the review of text messages constituted a search, and that the same rules that apply to a search of an employee's office apply equally to an intrusion into his electronic communications.[135] Further, the Court declined to decide which Fourth Amendment employment-based test from *O'Connor v. Ortega* applied—the plurality's "operational realities" test that looked at the specific facts of the employment situation on a case-by-case basis, or Justice Scalia's private employment equivalence test—because the Court decided the case on narrower grounds.[136]

The Court instead relied on the special needs exception to the warrant requirement, which holds that in certain limited instances a government employer need not get a warrant to conduct a search. When a government employer conducts a warrantless search for a "non-investigatory, work-related purpose," it does not violate the warrant requirement if it is "justified at its inception and if the measures are reasonably related to the objective of the search and not excessively intrusive in light of the circumstances giving rise to the search."[137] In the Court's judgment, the city had a "legitimate work-related rationale," and the scope of the search was reasonable and not "excessively intrusive."[138]

Like the city communication policy in *Quon*, as a condition of enrolling in EINSTEIN 2, each federal agency is required to enter into an agreement with DHS that certifies that certain log-on banners or computer user agreements are used to ensure employees are aware of and consent to the monitoring, interception, and search of their communications on federal systems.[139] Applying the "operational realities" test from *O'Connor*, the Department of

Justice's Office of Legal Counsel posits that use of the log-on banners on all federal computers will eliminate any expectation of privacy in communications transmitted over those systems.[140] Professor Orin Kerr takes a different approach, treating the terms of service of an Internet service contract—the equivalent to a log-on banner—as consent rather than an outright elimination of a reasonable expectation of privacy.[141] Under either approach, the conclusion reached is likely the same—the monitoring is in all likelihood reasonable.[142] However, *Quon* was limited to searches for a "noninvestigatory work-related purpose."[143] If EINSTEIN could be construed as overreaching this permissible purpose, say, by scanning e-mails for unlawful activity instead of simply malicious computer activity, a court may find its scope beyond *Quon*'s holding. Further, *Quon* insisted that these work-related investigations not be "excessively intrusive."[144] A reasonable argument could be made that monitoring the content of every employee communication is excessively intrusive. Additional questions remain. For instance, what is the scope of a non-investigatory, work-related purpose? Does scanning for malicious activity qualify as a work-related purpose? Does *United States v. Jones*'s physical intrusion test apply here where the employee's electronic *papers* and *effects* are being scanned?[145] Because no court has confronted a program like EINSTEIN, answers to these questions are unclear.

Monitoring Communications from Private Persons to Federal Employees

EINSTEIN not only monitors the computer activity of federal agency employees, but also any communications sent by a private person to a federal employee on his governmental e-mail or personal e-mail. One may argue that these concerns are more serious than in the employment context, on the theory that there is neither a presumption that an individual's privacy rights are diminished nor has the private actor consented to monitoring by clicking on a log-on banner or user agreement that would inform him of the privacy implications of his communication.

Some would argue that the third-party doctrine permits EINSTEIN's monitoring of private parties.[146] Traditionally, there has been no Fourth Amendment protection for information voluntarily conveyed to a third-party.[147] This doctrine dates back to the "secret agent" cases, in which any words uttered to another person, including a government agent or informant, were not covered by the Fourth Amendment.[148] Because federal employees have agreed to permit governmental monitoring of their communications, the Office of Legal Counsel (OLC) argues they are permitting *ex ante* surveillance of all their communications, including those from private persons to the federal employee's personal e-mail.[149]

However, the third-party cases have traditionally applied only to non-content information. In *Smith v. Maryland*, the Court noted that pen registers only disclose the telephone numbers dialed: "[n]either the purport of any communication between the caller and the recipient of the call, their identities, nor whether the call was even completed is disclosed by pen registers."[150] The case rested on the devices "limited capabilities."[151] The Ninth Circuit borrowed this reasoning in *Forrester*, where the panel distinguished "mere addressing" in an e-mail such as the to/from line, from "more content-rich information" such as the text in the body of an e-mail.[152] And as noted in *United States v. Warshak*, people still should expect privacy in the content of their telephone calls despite the ability of an operator to listen.[153] Further, the Supreme Court has noted that "the broad and unsuspected governmental incursions into conversational privacy which electronic surveillance entails necessitate the application of Fourth Amendment safeguards."[154] These cases severely diminish the argument that the third-doctrine permits absolute access to private communications. Instead, it could be reasonable to conclude from these cases that the third-party doctrine would permit access to the routing information of Internet communications, but might not go so far as to allow monitoring of the content of those communications.

Additionally, the OLC contends that under the "secret agent" cases the government can monitor private communications even if the sender is unaware that the recipient is a federal employee or did not anticipate that the communication would be opened on a federal computer.[155] The "secret agent" cases generally hold that "when a person communicates to third-party even on the understanding that the communication is confidential, he cannot object if the third party conveys that information or records thereof to law enforcement authorities."[156] Because these cases do not limit the instances this rule can be applied, it seems reasonable that they can be applied to EINSTEIN.

Alternative to Traditional Warrant Requirement

Assuming both federal employees and those communicating with them have a reasonable expectation of privacy in the contents of their communications, EINSTEIN must be tested under the general reasonableness requirement of the Fourth Amendment. A search is generally unreasonable without a warrant or some individualized suspicion.[157] However, under the "special needs exception" cases, the Court has held that when there are special governmental needs, beyond normal law enforcement, the government may need neither a warrant nor any level of individualized suspicion.[158] To determine whether the special needs exception applies, the Court balances the individual's privacy expectations against the governmental interest at stake.[159] This rule has been used to support certain police searches at checkpoints such as sobriety roadblocks,[160] border searches,[161] and checkpoints looking for a witness to a crime.[162] However, the Court did not permit a drug interdiction checkpoint when the "primary purpose was to detect evidence of ordinary criminal wrongdoing."[163]

Here, an argument could be made that the nature of cybersecurity and the impracticability of obtaining a warrant might justify application of the special needs doctrine to the EINSTEIN program.[164] The ostensible primary purpose of the program's cybersecurity measures is not for ordinary law enforcement needs, but instead to protect the critical infrastructure of the nation. Moreover, the government will need to act quickly if the program is to be feasible.[165] It could also be argued, however, that unless the threat required immediate review, a government agency should obtain a warrant based upon probable cause to review personally identifiable information, or, at a minimum, review the communications in a redacted format that includes only the threat information and no personally identifiable information.[166] As one commentator noted, it is nearly impossible to predict what is reasonable without knowing the severity of the cybersecurity threat and the exact measures taken to meet it.[167]

Privacy and Civil Liberties Oversight

In addition to the Fourth Amendment, there may be other mechanisms for protecting the privacy of Internet users. Indeed, the Constitution is only the floor for privacy protections. In many instances, Congress and state legislatures have created privacy protections beyond what is protected under their respective constitutions. These include statutes such as the Electronic Communications Privacy Act[168] and the Privacy Act of 1974.[169]

As to existing privacy protections, EINSTEIN has several privacy safeguards. For example, federal agencies are required to post notices on their websites that computer security information is being collected.[170] The computer programs recording network flow records strip down the information so that minimal content information is exposed.[171] Further, only the raw computer network traffic that contains malicious activity is viewed by DHS personnel; any "clean" traffic is promptly deleted from the system.[172]

Information is only collected when it relates to an actual cyber threat.[173] Analysts handling the monitored communications are given privacy training on an annual basis.[174] These privacy protections are handled internally within DHS.

Jack Goldsmith, former head of the Office of Legal Counsel, has proposed a system of four oversight mechanisms similar to the Foreign Intelligence Surveillance Court[175] to ensure the reasonableness of the searches under EINSTEIN: (1) independent *ex ante* scrutiny to ensure that the governmental procedures stay within their statutory authority; (2) privacy protections such as minimization procedures, also subject to *ex ante* judicial review; (3) *ex post* oversight mechanisms, in which the Attorney General and the Director of National Intelligence report to Congress every six months regarding privacy compliance and the inspectors general from each agency also report to Congress on a yearly basis; and (4) a sunset provision requiring Congress to reapprove the regime four years into operation.[176]

Others have proposed there be some form of independent oversight beyond DHS's privacy office.[177] Additionally, there are proposals that content of communications not be shared with law enforcement officials or used in any non-cyber crime investigation, unless the data was obtained as part of a legitimate cybersecurity threat.[178]

Legislation in the 112th Congress

This section provides a brief description of proposed cybersecurity legislation in the 112th Congress that include provisions regarding the EINSTEIN program with particular emphasis placed on the provisions that implicate the legal issues discussed above.

S. 2105, Cybersecurity Act of 2012

S. 2105, the Cybersecurity Act of 2012 was introduced on February 14, 2012. Title II of the bill would amend the Federal Information Security Management Act (FISMA) to codify many of the current executive agency network intrusion practices described above. Specifically, the bill would enact a new Section 3553 of Title 44 of the United States Code to provide the Secretary of DHS with the statutory authority to oversee all information security policies that will be binding on all federal agencies.[179] The Secretary would be given the explicit statutory authority to "acquire, intercept, retain, use, and disclose communications and other system traffic that are transiting to or from or store on agency information systems and deploy countermeasures with regard to the communications and system traffic." In light of the test phase of EINSTEIN 3, this section apparently would vest in the Secretary the authority to not only intercept communications, but to react to actionable data based on perceived threats. Consistent with current practice, The Cybersecurity Act of 2012 would leave the Department of Defense (DOD), the Central Intelligence Agency (CIA), and the Office of the Director of National Intelligence in charge of their respective network intrusion systems. This bill would also permit DHS to contract with private ISPs to further its mission of preventing network intrusions.

Additionally, the National Center for Cybersecurity and Communications (NCCC) created under Title III of the Cybersecurity Act of 2012[180] would have various responsibilities with respect to federal networks. First, the NCCC must implement and enforce a security system for the federal information infrastructure,[181] which does not include national security systems or information systems used by the Department of Defense, the military or the intelligence community.[182] Second, the Center is to provide cybersecurity technology, upon request, and information, both classified and unclassified, to owners of national information infrastructure. Third, the NCCC must develop a national incident response plan and is responsible for coordinating national cyber incident response efforts. Finally, the NCCC must create an information sharing system that collects information from and redistributes information to federal agencies, state and local governments, national information infrastructure, critical infrastructure, and the private sector. Both federal agencies and critical infrastructure have an affirmative obligation to provide certain information to the Center's information sharing program.[183] Other entities, including state and local governments and private sector actors, can participate voluntarily in the program.[184]

Title III instructs the NCCC to "develop, oversee the implementation of, and enforce policies, principles, and guidelines on information security for the federal information infrastructure, including exercise of the authorities under the Federal Information Security Management Act of 2002."[185] Therefore, it appears that NCCC is intended to exercise the authority granted to the Secretary in Title II.

Several provisions in Title III also appear to overlap with authority granted to DHS under FISMA in Title II. For example, both Title II and Title III require risk assessments to be completed on agency information systems. In Title III, NCCC must conduct risk assessments on all federal information systems, which excludes national security systems and systems used by the

Department of Defense, the military, or the Intelligence Community.[186] In Title II, DHS is instructed to conduct risk assessments on agency information systems, except national security systems and systems used by the Department of Defense, the Central Intelligence Agency (CIA), and the Office of the Director of National Intelligence (DNI).[187]

Additionally, the security policies and procedures created under Title II and Title III appear to apply to different sets of federal agencies because of different exemptions granted under each Title. While national security systems and systems used by the Department of Defense, CIA, and DNI are exempted from the security policies and procedures implemented by DHS pursuant to Title II, other members of the intelligence community are not exempted.[188] In contrast, all national security systems and systems used by the Department of Defense, the military, and the entire Intelligence Community are exempted from the requirements NCCC must implement under Title III.[189] Therefore, it appears that the NCCC may be implementing two different sets of policies for two different sets of federal agencies, one set under Title II and one set under Title III.

It appears that all federal agencies must participate in the NCCC information sharing system, including complying with the affirmative obligation to provide information to the Center.[190]

H.R. 3674, Promoting and Enhancing Cybersecurity and Information Sharing Effectiveness Act of 2012 (PRECISE Act)

H.R. 3674, the Promoting and Enhancing Cybersecurity and Information Sharing Effectiveness Act of 2012 (PRECISE Act) was introduced on December 15, 2011. As introduced, H.R. 3674 did not expressly address federal efforts to implement and strengthen the EINSTEIN program. However, on April 18, 2012, the House Committee on Homeland Security held a mark up of the bill and agreed to an amendment that would codify DHS's EINSTEIN program.

H.R. 3674, as amended, would expressly permit DHS to "acquire, intercept, retain, use, and disclose communications" coming to or from any federal computer and to deploy countermeasures against potential malicious activity.[191] This is a codification of EINSTEIN 3's interception and response system tested on one agency by DHS in 2010.[192] Under this bill, an interception or countermeasure[193] must be "reasonably necessary" for protecting the federal system against a cybersecurity threat.[194] DHS may only retain the information to protect the federal systems against attack or to mitigate future attacks. The intercepted information may be used for law enforcement purposes, but only if authorized by the Attorney General.

Under this bill, notice must be provided to users of federal systems concerning the potential interception or use of their communication. Generally, this will take the form of a log-on banner for federal employees[195] or a federal website privacy policy page.[196] What is unclear, however, is whether the disclaimer provision covers situations where a private person is communicating with a federal employee. In many instances, that person may not be informed by the federal agency that their communication is being monitored. As just discussed, the third-party doctrine may vitiate any Fourth Amendment protection in this communication such that a notice to a private person communicating with a federal computer is not necessary.[197] But again, the third party-doctrine has primarily been applied to non-content information.[198] H.R. 3674 does not distinguish between content and non-content communications, but permits interception of any communication that is "associated with a known or reasonably suspected cybersecurity threat."[199] As such, the Fourth Amendment may be applicable to the content-based information collected under the procedures established by H.R. 3674.

Additionally, DHS is permitted to contract with private entities that provide cybersecurity services in order to facilitate the interception and use of communications coming to and from federal computers.

H.R. 4257, Federal Information Security Amendments Act of 2012

H.R. 4257, the Federal Information Security Amendments Act of 2012 was introduced by Representatives Daryl Issa and Elijah Cummings on March 26, 2012. This bill would amend the Federal Information Security Management Act of 2002 (FISMA) to require all federal agencies to continuously monitor their computer networks for malicious activity—a codification of the EINSTEIN program—and would impose additional cybersecurity requirements of all federal agencies.[200] Unlike the Lieberman-Feingold or Lungren bills, H.R. 4257 does not put control in the hands of DHS. Instead, the Office of Management and Budget (OMB) would oversee the implementation and requirements of information security policies.[201] In conjunction with the National Institute of Standards and Technology (NIST), OMB would create standards and guidelines for cybersecurity protection. The Director of OMB would be charged with ensuring agency compliance with these standards.

Under H.R. 4257, each federal agency would be responsible for "providing information security protections commensurate with the risk and magnitude of the harm resulting from unauthorized access, use, disclosure, disruption, modification or destruction of—(i) information collected or maintained by or on behalf of an agency; and (ii) information infrastructure used or operated by an agency or by a contractor of an agency or other organization on behalf of an agency; ..."[202] Each agency would be required to establish a chief information officer responsible for developing and overseeing the agencies information security program.

Additionally, OMB would be charged with establishing a federal information security incident center, which would act as a hub for federal cybersecurity systems. This center would, among other duties, provide assistance to agencies, compile and analyze information about cybersecurity incidents, and inform agencies about known cyber threats.[203]

Absent from H.R. 4257 are any express privacy or civil liberties protections. Potentially, these protections may be established in future guidelines created by NIST and OMB, but this is unclear from the language of the bill. Further, unlike H.R. 3674 (Lungren), there is no requirement that computer users be informed that their communications with a federal computer can be monitored. Absent from H.R. 4257 are any express privacy or civil liberties protections.

LEGAL ISSUES RELATED TO CYBERSECURITY THREAT INFORMATION SHARING

Many policymakers have argued that there is a need for the federal government and owners and operators of the nation's critical infrastructures to share information on vulnerabilities and threats, and to promote information sharing between the private and public sectors in order to protect critical assets from cybersecurity threats. Private sector entities may wish to share information with one another about threats they have faced or are currently facing. They may also wish to collaborate in devising solutions to these security issues. Additionally, the government may have information about cybersecurity threats that would be similarly useful to potential targets in the private sector. The government may also see value in having access to information from the private sector about cybersecurity threats.

Obstacles to information sharing may exist in current laws protecting electronic communications or in antitrust law. The Fourth Amendment, the Telecommunications Act of 1934, and state laws may also affect the legality of information sharing by the private sector. Entities that share information may also be concerned that sharing or receiving such information may lead to civil and criminal liability, or that shared information may contain proprietary or confidential information that could be disclosed to competitors or government regulators.

Electronic Communications Privacy Act[204]

The Electronic Communications Privacy Act (ECPA) generally prohibits (1) the interception of wire, oral or electronic communications (wiretapping);[205] (2) access to the content of stored electronic communications and to communications transaction records;[206] and (3) the use of trap and trace devices and pen registers.[207]

ECPA generally prohibits intercepting wire, oral, or electronic communications by means of an electronic, mechanical or other device, but sets forth a number of exceptions to the general prohibition.[208] Relevant to this discussion, ECPA provides a general exemption for communications service providers, permitting them to intercept communications when incidental to "the rendition of service or the protection of the rights or the property of the provider of that service," or protecting themselves against fraud.[209] This exemption does not apply to random monitoring except where used for mechanical or service quality control checks. Communications service providers are also permitted to intercept communications in order to assist federal and state officials operating under a judicially supervised interception order,[210] and for the regulatory activities of the Federal Communications Commission.[211] In addition, communications service providers are permitted to intercept communications with customer consent.[212]

Under the stored communications provisions of ECPA, providers of electronic communication services (ECS) to the public may not disclose the contents of any "communication while in electronic storage by that service."[213] Public remote computer service (RCS) providers similarly may not disclose the contents of

> any communication which is carried or maintained on that service – (A) on behalf of, and received by means of electronic transmission from (or created by means of computer processing of communications received by means of electronic transmission from), a subscriber or customer of such service; (B) solely for the purpose of providing storage or computer processing services to

such subscriber or customer, if the provider is not authorized to access the contents of any such communications for purposes of providing any services other than storage or computer processing.[214]

Both ECS and RCS providers may not disclose any "record or other information pertaining to a subscriber to or customer of such service (not including the contents of communications covered by [the disclosure restrictions described above]) to any government entity."[215]

However, the statute does provide a number of exceptions under which an ECS or RCS provider may disclose the contents of a communication. These exceptions cover disclosures made

- to the addressee or intended recipient of the communication;
- with the consent of the sender, addressee, or intended recipient of the communication, or to the subscriber in the case of remote computing service;
- in order to forward such communication to its destination;
- as may be necessarily incident to the rendition of the service or *to the protection of the rights or property of the service provider;*
- to the National Center for Missing and Exploited Children;
- to law enforcement if the contents were inadvertently obtained by the service provider and appear to pertain to the commission of a crime; and
- to a government entity, if the provider, in good faith, believes that an emergency involving danger of death or serious physical injury to any person requires disclosure.[216]

With respect to pen registers and trap and trace devices, ECPA outlaws installation or use of a pen register or trap and trace device, except under one of seven circumstances:

- pursuant to a court order issued under sections 3121-3127 (pen registers and trap and trace devices);
- pursuant to a Foreign Intelligence Surveillance Act (FISA) court order;[217]
- with the consent of the user;
- when incidental to service;
- when necessary to protect users from abuse of service;
- when necessary to protect providers from abuse of service;[218] or

- in an emergency situation.[219]

Some have argued that the framework provided by ECPA may be an obstacle to sharing cyber threat information among communications service providers or between such entities and the government,[220] and may prevent them from acting to protect their customers and networks. The statute permits service providers to conduct random monitoring of communications in order to perform mechanical or service quality control checks; however, these purposes may not sufficiently capture the wholesale monitoring of networks to detect or intercept cyber threats.[221] Additionally, the restrictions on voluntary disclosures of the contents of communications and addressing information are generally limited to the purpose of protecting the service provider's rights or property. Consequently, ECPA may hinder sharing of information about cyber threats where the service provider is not the target of the threat. Given this uncertainty, providers may be hesitant to share cyber threat information as violating ECPA can expose them to criminal penalties and private civil liability.

Antitrust Law

Companies may be assisted in combating cybersecurity threats by sharing information with one another about threats they have faced or are currently facing. Companies may also wish to collaborate in devising solutions to these security issues. The antitrust laws are often cited as an impediment to such collaboration. This is so because if a collaboration is found to violate antitrust laws, the collaborating entities may be subject to civil and criminal penalties.[222]

Section 1 of the Sherman Antitrust Act prohibits contracts, combinations, and conspiracies in restraint of trade.[223] The Supreme Court has found that not all contracts or combinations that restrain trade are forbidden by the Sherman Act; rather, only those agreements that unreasonably restrain trade are prohibited.[224]

Nonetheless, when competitors share information with one another, concerns regarding violations of the antitrust laws may arise.[225] The sharing of information may create the opportunity to conspire to fix prices, restrain output, or otherwise agree to unreasonably restrain competition to the detriment of consumers.

Two types of analyses are used to determine the lawfulness of collaborative activity among competitors: per se and rule of reason.[226] The per se analysis is applied to collaborations that have been found to be always or almost always in violation of the antitrust laws because they result in raising prices or reducing output without any appreciable benefit to competition.[227] Only the most egregious collaborations, such as those to fix prices, rig bids, or reduce output, are considered to be per se illegal.[228] All other collaborations among competitors are subject to review under the rule of reason standard.[229] The rule of reason consists of a flexible inquiry into the potential competitive benefits of an agreement as they are weighed against the potential competitive harms. Most agreements to share information will likely be reviewed under the rule of reason standard.[230] Most collaborations among competitors that exist for the sole purpose of combating cybersecurity threats would be analyzed under the rule of reason standard.

Collaboration among competitors may include a wide variety of activity including research and development, shared manufacturing facilities, and other joint ventures.[231] Agreements to share information may be a part of other broader collaborative activities, or an end unto themselves. The Department of Justice (DOJ), and the Federal Trade Commission (FTC) recognize that information sharing among competitors often has pro-competitive and efficiency-enhancing benefits that may outweigh any anticompetitive risks.[232] The DOJ and the FTC, therefore, have devised guidelines to aid companies in developing collaborative business plans that minimize antitrust concerns.[233] The first aspect of the agreement that the agencies will examine is the extent of the collaboration and the purpose for the collaboration.[234] To the extent that the sharing of information is limited to the purpose of aiding in combating cybersecurity threats, it is likely that the antitrust concerns raised by any potential agreement would be limited as well.[235]

Groups of competitors wishing to collaborate to combat cybersecurity threats, even when following the DOJ and FTC's guidelines, may nonetheless be concerned about antitrust scrutiny. To aid these groups, the DOJ has developed a process for the groups to submit their plans to collaborate to the DOJ for a determination by the agency of whether the proposed collaboration would raise antitrust concerns.[236] It is called the Business Review Procedure. The procedure has been used in the cybersecurity context in the past. For example, the Electric Power Research Institute (EPRI) requested that the DOJ review its proposal to share information related to cyber threats. After examining the proposal, the DOJ issued a business review letter stating that the DOJ was not inclined to initiate an antitrust enforcement action against the collaborative efforts of EPRI on the theory that the proposal would reduce cybersecurity costs and may have a pro-competitive effect.[237] Nonetheless, the DOJ, as it always does in these circumstances, reserved the right to pursue any antitrust concerns should the collaborative effort prove to have a future anticompetitive effect.

Liability for Information Sharing

Some have argued that sharing or receiving information about cybersecurity threats could potentially expose private sector entities to increased liability. To the extent that ECPA, antitrust laws, or other federal or state laws prohibit private sector entities from sharing cybersecurity threat information amongst themselves or with the government, violating these laws could lead to civil or criminal penalties imposed by the government.[238] Additionally, both ECPA and the antitrust laws provide private rights of action for harmed parties to recover damages from entities that have violated these statutes.[239] Consequently, violating ECPA or the antitrust laws may also expose entities to private civil liability.

Concerns about private civil liability for information sharing may also arise based on the effect that information sharing may have on private civil actions based on injuries caused by a defendant's negligent actions. One way of proving negligence is by convincing a jury that the defendant did not act reasonably in the face of a foreseeable risk.[240] In the absence of a foreseeable risk, a defendant typically has no judicially enforceable duty to mitigate that risk.[241] However, if a defendant has received information about an active cybersecurity threat, then that would tend to show that the risk of attack from such threat was a foreseeable one. In other words, notice of cybersecurity risks might lead a jury to find that the defendant had a duty to act reasonably. For example, if a defendant is using software package X in its information infrastructure, and the defendant receives information from other private sector entities or the government that software package X has been vulnerable to cyberattacks, the receipt of this information may lead a jury to conclude that the defendant was aware of the risk presented by using that software package. If such a duty were found, then the defendant could be liable for any harm that resulted from its negligence.

Receiving information about cybersecurity threats may also be relevant to whether the actions taken by a defendant in the face of a foreseeable risk were reasonable. In order to determine whether a defendant's actions were reasonable, juries are often asked to balance the foreseeable risks of the defendant's actions with the foreseeable risks of the defendant's inaction.[242] For example, shared cybersecurity threat information may include effective and low-cost measures that could be taken to mitigate or prevent a threat. A jury evaluating whether a defendant had acted negligently may find the fact that the defendant had knowledge of effective and low-cost preventative measures may determine that the defendant should be held to a higher standard of care than if the defendant had not received such information.[243]

Protection of Proprietary or Confidential Business Information[244]

Sharing cybersecurity threat information may raise concerns about how that information would be used. For example, there may be concerns that other businesses could use the information to gain a competitive advantage. There may also be concerns that cybersecurity threat information shared with the government might be used for regulatory purposes unrelated to cybersecurity. As a result, some private sector entities may be hesitant to voluntarily share cybersecurity-related information with other businesses or with the government.

For example, voluntary sharing of cybersecurity threat information with the government may be inhibited by concerns that such information might be made publicly available under the Freedom of Information Act of 1974 (FOIA), which regulates the disclosure of agency records held by the federal government.[245] Other potential obstacles to sharing information with the government are agency rules or judicial doctrine regarding ex parte communications, the rules of discovery in civil litigation, and state open records laws requiring public disclosure.

Information that is designated as critical infrastructure information (CII) under the Critical Infrastructure Information Act (CIIA) is protected from disclosure under FOIA. Similarly, the CIIA provides that CII will not be subject to agency rules or judicial doctrine regarding ex parte communications. With respect to concerns about litigation, CIIA limits the use of CII in civil litigation and provides that sharing CII with the agency does not count as the "waiver of any applicable privilege or protection provided under law," such as trade secret protection or the attorney-client privilege.[246] CIIA also authorizes the use or disclosure of such information by officers and employees in furtherance of the investigation or the prosecution of a criminal act; or for disclosure to Congress or the Government Accountability Office.

Many of these concerns are also raised in the context of protecting information collected from critical infrastructure, and are discussed in more detail *supra* at "Freedom of Information."

Privacy and Civil Liberties

Privacy and civil liberties advocates argue that several of the proposed cybersecurity information sharing measures go too far in eroding privacy protections.[247] For instance, CISPA and the SECURE IT Act permit private sector use of cybersecurity systems and sharing of cyber threat information "[n]otwithstanding any other provision of law."[248] This provision has the effect of overriding privacy protections such as ECPA and the Privacy Act of 1974. One commentator noted that although some changes are necessary to authorize cyber activities, a broad exclusion of these laws in the cybersecurity area would be "inconsistent with the promise of privacy that undergirds the Wiretap Act and the SCA."[249]

There is also concern among privacy and civil liberties groups that defense agencies like the National Security Agency (NSA) would have access to Internet information obtained through cybersecurity information sharing programs. Generally, defense agencies are not employed in the domestic law enforcement arena.[250] These groups warn that defense agencies like the NSA are not subject to the same oversight and transparency as civilian agencies such as DHS.[251] Observers point to its warrantless wiretapping program in 2001 as proof that the NSA should not be given control over monitoring of domestic Internet activity.[252] These advocates suggest that any proposed information sharing plan clearly state which civilian agencies will have access to this information.[253] This would prevent, in their view, the NSA or other military agencies from inadvertently getting access to this data.

Legislation in the 112th Congress

This section provides a brief description of proposed cybersecurity legislation in the 112th Congress that include regulatory provisions addressing the sharing of cybersecurity threat information amongst the private sector and between the government and the private sector. Particular emphasis has been placed on the provisions that implicate the legal issues discussed above.

H.R. 3523, Cyber Intelligence Sharing and Protection Act of 2011, as Reported

H.R. 3523, the Cyber Intelligence Sharing and Protection Act of 2011 (CISPA), was introduced on November 30, 2011, by Representative Rogers of Michigan, to facilitate sharing of cyber threat intelligence information between the Intelligence Community (IC)[254] and the private sector. On December 1, 2011, the House Permanent Select Committee on Intelligence (HPSCI) held a mark up of the bill and favorably ordered reported an amended version to the House.[255] On April 16, 2012, the HPSCI released a discussion draft with new proposed amendments that would be considered when the bill goes to the floor.[256] On April 17, H.R. 3523 was reported with an amendment.[257] The HPSCI released "H.R. 3523–Draft Amendment in the Nature of a Substitute provided to the House Rules Committee–April 19, 2012.[258]

As reported out of the Rules Committee, CISPA would direct the Director of National Intelligence (DNI) to establish procedures under which IC elements would be allowed to share cyber threat intelligence with the private sector and utilities and to encourage sharing.[259] CISPA defines "cyber threat information" as "information directly pertaining to a vulnerability of, or threat to a system or network of a government or private entity, including ... (A) efforts to degrade, disrupt or destroy such system or network; or (B) efforts to gain unauthorized access to a system or network...."[260] Classified cyber threat intelligence information from the IC would only be shareable with entities or persons with appropriate security clearances, and only if sharing would be consistent with the need to protect national security.[261] Entities receiving shared cyber threat intelligence from the IC would be required to take measures to protect that information from unauthorized disclosure.[262] CISPA also directs the DNI to issue guidelines allowing heads of IC elements to grant clearances through an expedited process to employees or officers of certified entities and to certified entities.[263] CISPA explicitly provides that it does not create any right or benefit to cyber threat information as a result of the government's sharing of such information with any public sector entity or utility.

To respond to criticisms about the potential for redisclosure of cyber threat information by recipients, CISPA provides that an entity that receives cyber threat intelligence would be prohibited from further disclosure of such information to an entity other than the federal government or a certified entity, notwithstanding any other provision of law.

In order to address the perceived legal obstacles to information sharing among the private sector, CISPA would give private sector entities the explicit authority to use cybersecurity systems to identify and obtain cybersecurity threat information to protect their rights and property, and to share such information with any other entity, including with the federal government, "notwithstanding any provision of law."[264] Entities that provide cybersecurity goods and services, also known as cybersecurity providers, would have similar authority with the consent of their customer to use cybersecurity systems to identify, obtain, and share cyber threat information from the networks of consenting customers, with any other entity designated by the customer including the federal government if designated.[265] The "notwithstanding any provision of law" language means that information sharing is authorized even if another federal or state privacy law, or other law, would protect the information against disclosure.[266] CISPA requires any federal agency or department in receipt of cyber threat information to provide that information to the DHS which is authorized upon request to share such information with another federal agency or department.

The bill would require cyber threat information to be shared subject to restrictions imposed by the submitting entity, including appropriate anonymization or minimization of such information.[267] CISPA would also explicitly prohibit shared information from being used by businesses to gain an unfair competitive advantage to the detriment of the entity sharing the information.[268] CISPA also explicitly exempts information shared with the federal government from disclosure under FOIA and under state or local public disclosure laws.[269] Additionally, the bill would provide that information shared with the federal government be considered proprietary information that could not be disclosed to an entity outside of the federal government, except with the authorization of the entity sharing the information, and could not be used by the government for regulatory purposes.[270] In addition, such information may not be shared by the federal government if it would undermine the purpose for which it was shared; or the federal government recipient determines, unless directed otherwise by the President, that providing the information would undermine the purpose of sharing. The bill also provides that information shared with the federal government be handled consistent with the protection of sources and methods and with national security.

With respect to concerns about increased liability resulting from sharing cybersecurity threat information, CISPA would provide civil and criminal immunity for entities that, acting in good faith, use cybersecurity systems, share information, or make decisions based on such information in accordance with the new authorities created by the bill.[271]

The federal government would be permitted to use shared information for any lawful non-regulatory purpose, provided that at least one significant purpose of the use is a cybersecurity purpose or protection of national security.[272] The federal government is also specifically prohibited from searching shared cyber threat information unless it is for the purpose of cybersecurity or protection of national security.[273]

CISPA includes an anti-tasking restriction that explicitly prohibits the government from conditioning the sharing of cyber threat intelligence on the sharing of private sector information with the government.[274]

For the government's intentional or willful violations of the provisions concerning the disclosure, use, or protection of voluntarily shared cyber threat information, CISPA would provide liability for actual damages or $1,000, whichever is greater, plus reasonable attorney fees, and costs.

With respect to concerns about transparency and oversight, CISPA would require an annual unclassified report (with classified annex) by the Intelligence Community Inspector General on the type and use of information shared with the federal government, including actions taken and impacts on privacy and civil liberties.

CISPA would preempt any state or local law that restricts or regulates the use of cybersecurity systems and the sharing of cyber threat information. CISPA would not affect any other authorities for use of a cybersecurity system or to identify, share, or obtain cyber threat intelligence. CISPA does not alter DOD, NSA, or the intelligence communities authority to direct private sector or government cybersecurity efforts. In addition, CISPA does not change existing information sharing relationships, prohibit or require new ones, or modify the government's authority to protect sources and methods and national security.

The Director of National Intelligence must establish and issue procedures and guidelines, in consultation with the Secretary of Homeland Security, to ensure that such procedures and such guidelines permit the owners and operators of critical infrastructure to receive all appropriate cyber threat intelligence in the possession of the federal government; and expeditiously distribute such procedures and such guidelines to departments and agencies of the federal government, private-sector entities, and utilities.

H.R. 3674, PRECISE Act

H.R. 3674, the Promoting and Enhancing Cybersecurity and Information Sharing Effectiveness Act of 2011 (PRECISE Act), was introduced on December 15, 2011, by Representative Lungren. On February 1, 2012, the Subcommittee on Cybersecurity, Infrastructure Protection, and Security Technologies of the House Homeland Security Committee held a mark up of the bill, and it was forwarded to the full committee by voice vote. On April 18, 2012, full committee consideration and mark up was held on a substitute bill that changed the information sharing and other provisions. The measure was reported favorably to the full House (as amended) 16-13.[275]

Section 3 of the PRECISE Act, as reported, would require the Department of Homeland Security to make cyber threat information and other information in DHS's possession available to critical infrastructure owners and operators in a manner that is consistent with statutory and other restrictions on the dissemination of such information.

In addition, section 3 would establish in DHS a National Cybersecurity and Communications Integration Center (NCCIC) for sharing cyber threat information and exchanging technical advice, and assistance. The majority of the members of the advisory board for NCCIC will be from critical infrastructure sectors. NCCIC's charter will be developed by DHS. The NCCIC Board will submit a report to Congress annually on NCCIC's information sharing activities. The DHS is authorized to issue warnings on cybersecurity threats. The DHS may not disclose information that is the basis of the warning without the express consent of the entity that submitted the information. In addition, DHS may not disclose information that is proprietary, business sensitive, relates to the submitting entity, or is not appropriate for public disclosure; and any information restricted by statute, rule, or regulation, and information relating to sources and methods and national security.

S. 2102, Cybersecurity Information Sharing Act of 2012

S. 2102, the Cybersecurity Information Sharing Act of 2012 (CISA), was introduced on February 13, 2012, by Senator Feinstein, for the purpose of improving the sharing of cybersecurity information among entities in the private sector, and between the private sector and the government. The provisions of CISA have also been incorporated, largely without change, into Title VII of S. 2105, the Cybersecurity Act of 2012, as introduced on February 14, 2012.

CISA would address the perceived obstacles to information sharing in existing law by giving private entities affirmative authority to monitor their own information systems for cybersecurity threats, or the information systems of a consenting third party.[276] CISA would also provide explicit authority for private entities to disclose and receive lawfully obtained cybersecurity information so long as the shared information is used for cybersecurity protection and reasonable efforts are made to safeguard individually identifiable information.[277] However, nothing in CISA would be permitted to be construed to authorize price fixing or market allocation between competitors.[278]

Under CISA, the Secretary of DHS would be authorized to designate cybersecurity exchanges, for the purpose of efficiently receiving and distributing cybersecurity threat indicators.[279] Nonfederal entities are explicitly given the authority to provide cybersecurity threat indicators to a cybersecurity exchange, which may only use, retain, or further disclose shared information for the purpose of protecting against or mitigating cybersecurity threats.[280]

CISA would provide that information shared with a cybersecurity exchange would be exempt from disclosure under FOIA, as well as any restrictions on *ex parte* communications.[281] Sharing information with a cybersecurity exchange would not constitute a waiver of any applicable privilege regarding the information, including any trade secret protection.[282] Furthermore, no federal entity would be permitted to use a cybersecurity threat indicator as evidence for a regulatory enforcement action against the entity that shared the information.[283]

Federal entities would not be permitted to disclose cybersecurity threat information unless the disclosure is made to protect a federal entity from a cybersecurity threat, or to mitigate a cybersecurity threat to another component, officer, employee, or agent of the federal entity with cybersecurity responsibilities, any cybersecurity exchange, or a private entity that provides a federal entity with an electronic communication service, remote computing service, or cybersecurity service.[284] The recipient of information from a federal entity must also comply with any requirements regarding the protection and further disclosure of such information.[285]

Additional restrictions would apply if cybersecurity threat information was to be shared with law enforcement. Federal cybersecurity exchanges could only disclose information to law enforcement if the information appears to relate to a crime which has been, is being, or is about to be committed, and if minimization procedures developed by the Secretary and approved by the Attorney General permit such disclosure.[286] Federal entities that are not cybersecurity exchanges may use cyber threat information to protect against cybersecurity threats, but must comply with similar restrictions on disclosing shared information to law enforcement.[287] Any disclosure of cyber threat information to a nonfederal entity shall be accompanied by a written agreement under which the recipient of the information agrees that the information will only be used in a manner consistent with the restrictions on disclosures to law enforcement.[288] CISA directs the Secretary of DHS to devise minimization procedures to protect individually identifiable information from unnecessary disclosure.[289] These procedures are to be developed in consultation with privacy and civil liberties experts, the Director of National Intelligence, and the Secretary of Defense.

CISA provides immunity from civil and criminal liability arising from monitoring activities or voluntary disclosure of cyber threat information in compliance with CISA. However, this immunity only applies if the disclosure is made (1) to a cybersecurity exchange, (2) by a provider of cybersecurity services to a customer, (3) to a private entity or governmental entity that provides or manages critical infrastructure, or (4) to any other private entity if the threat information is also provided to a cybersecurity exchange within a reasonable amount of time.[290] Immunity would also be provided if an entity acts in good faith reliance that such actions are permitted by CISA.[291] No liability protections would attach to conduct that knowingly and willfully violates CISA.[292] With respect to negligence based actions, CISA would bar civil or criminal liability based on the reasonable failure to act on information received. No breach of contract claims could be brought based on compliance with lawful restrictions placed on shared information.[293] However, none of these protections could be construed to limit liability for a failure to comply with the requirements imposed on the use and protection of information.[294]

S. 2105, Cybersecurity Act of 2012

Title VII of the S. 2105, the Cybersecurity Act of 2012, contains virtually the same provisions as S. 2102, the Cybersecurity Information Sharing Act of 2012, discussed above. While Title VII of S. 2105 authorizes the Secretary of Homeland Security to designate both federal and nonfederal entities as cybersecurity exchanges, with the goal of "efficiently receiv[ing] and distribut[ing] cybersecurity threat indicators...."[295] Title III of the bill would also address this need by directing the new National Center for Cybersecurity and Communications (NCCC or Center) to create its own information sharing program.[296] Specifically, the NCCC would be charged with creating an information sharing system that collects information from and redistributes information to federal agencies, state and local governments, national information infrastructure, critical infrastructure, and the private sector. Both federal agencies and critical infrastructure would have an affirmative obligation to provide certain information to the Center's information sharing program.[297] Other entities, including state and local governments and private sector actors, would be permitted to participate voluntarily in the program.[298]

It may not be clear how these separate authorities would interact. For example, under Title VII, DHS would be required to designate a lead cybersecurity exchange within 60 days of the enactment of the act.[299] Following this interim period, which can only last 60 days, it is unclear if the NCCC program will be designated as a cybersecurity exchange. If the NCCC program is designated as an exchange, either the lead or an additional federal exchange,[300] the restrictions and protections outlined in Title VII would likely apply. Title VII creates specific limitations on the use of information in the exchange by federal entities,[301] nonfederal entities,[302] and the exchange itself.[303] Additionally, information shared in an exchange is explicitly exempted from Freedom of Information Act requests[304] and ex parte communications limitations.[305] Title VII also provides immunity from liability based on lawfully obtained cybersecurity information that is voluntarily disclosed to an exchange.[306] Finally, exchanges are bound by specific requirements regarding with whom an exchange can share classified information, including restricting access to people with "an appropriate security clearance."[307]

However, if the NCCC program is not designated as an exchange, it arguably appears that Title VII would not apply to the program at all. Title III includes far fewer restrictions on the use of the information gathered by the Center's program. Notably, Title III provides the Center much greater discretion in disseminating classified information, only instructing the Director to create procedures to ensure classified information is "appropriately shared between and among appropriate Federal and non-Federal entities...."[308] However, unlike Title VII exchanges, Title III does not specifically restrict who can receive classified information from the NCCC information sharing program. Furthermore, Title III does not provide protection from liability for entities that provide information to the NCCC program, even though certain entities, like covered critical infrastructure, are required to disclose incident-related information.[309]

S. 2151, SECURE IT Act

S. 2151, the Strengthening and Enhancing Cybersecurity by Using Research, Education, Information, and Technology Act of 2012 (SECURE IT Act), was introduced by Senator McCain on March 1, 2012. Title I of the

SECURE IT Act addresses the sharing of cybersecurity threat information among the private sector and between the private sector and the public sector.

Notwithstanding any other provision of law, the SECURE IT Act would give private sector entities the explicit authority to employ countermeasures and use cybersecurity systems to identify, obtain, or otherwise possess cyber threat information for the purpose of preventing, investigating, or otherwise mitigating threats to information security.[310] This authority applies to networks owned by an entity, or to other networks as authorized by the other networks' owner.[311] Entities would also be authorized to disclose cyber threat information to a cybersecurity center[312] or to any other entity for the same purposes.[313] Entities that provide information security products or services, also known as information security providers, would also be permitted to obtain, identify, possess, or disclose cyber threat information encountered in the course of providing such services. However, customers must be given a reasonable opportunity to authorize or prevent any disclosure, or to request anonymization of such information.[314]

Private entities would be permitted to share information directly with each other, but recipients of cybersecurity threat information would be obligated to comply with restrictions (such as anonymization) set forth by the entity providing the information.[315] Shared information may not be used to obtain an unfair competitive advantage. The act of sharing would not be considered a violation of the antitrust laws if shared to assist with information security.[316]

Providers of electronic communication services, remote computing services, or cybersecurity services to a federal agency or department would be required to provide any cyber threat information related to the provision of such services that is in the provider's possession.[317] The provider is also permitted to provide this threat information to a cybersecurity center,[318] and the notified federal department or agency is required to provide the threat information with a cybersecurity center.[319] A cybersecurity center would be required to share information with other cybersecurity centers,[320] and may disclose such information to other federal entities for cybersecurity or national security purposes, or for the prevention, investigation, or prosecution of any of the crimes that are eligible for an interception order under ECPA.[321] Such information may also be disclosed to a provider of electronic communication services, remote computing services, or cybersecurity services, for purposes related to those services.[322]

The SECURE IT Act would place restrictions on how information shared with cybersecurity centers could be disclosed or used.[323] Except for the disclosures provided in the preceding paragraph, information shared with a cybersecurity center could not be disclosed by the cybersecurity center without the consent of the entity that provided the information.[324] Information shared with a cybersecurity center would also be exempt from disclosure under FOIA and similar state or local laws requiring disclosure.[325] Shared information would also not be subject to any restrictions on *ex parte* communications.[326] Federal, state, tribal, and local governments would be prohibited from using or disclosing shared information for regulatory purposes.[327]

In order to address concerns that private sector entities may have about liability, the SECURE IT Act provides civil and criminal immunity for actions authorized under the bill.[328] Additionally, no cause of action would be permitted against an entity for using, receiving, or disclosing cyber threat information, or for any act or omission following the lawful receipt of such information.[329] Notwithstanding these provisions, no immunity would be provided for unlawful disclosures of classified information.[330]

PREEMPTION

As the body of federal cybersecurity law grows, the possibility that it will preempt conflicting state law will increase with it. After September 11, 2001, states took various measures to protect their critical infrastructure. This included defining "critical infrastructure," creating security standards for these entities, and carving out exceptions under public disclosure laws so vital information would not get into the hands of bad actors.

It is well established that the Supremacy Clause of the United States Constitution can invalidate any state law that interferes with or is contrary to federal law.[331] This is known as preemption. The preemptive effect of a federal statute can be either expressly stated in the statute or implied by the structure and purpose of the legislation.[332] If there is express language, the court will interpret the words used by Congress and assume that the ordinary meaning of the text expresses the legislative purpose.[333] For example, if Congress uses broad language in its preemption provision, the court will construe its preemptive effect broadly.[334] Absent explicit preemptive language, there are two types of implied preemption: (1) field preemption, where the federal regime is "so pervasive to make the reasonable inference that Congress left no room for the States to supplement it";[335] and (2) conflict preemption, where state law "stands as an obstacle to the accomplishment and execution of the full purposes and objectives of Congress."[336]

Because any preemption analysis relies on congressional intent, the language of the statute is of primary importance. Many of the proposals provide explicit language preempting state laws. For example, Title I of S. 2105, the Cybersecurity Act of 2012, contains an express preemption provision, stating: "This Act shall supersede any statute, provision of a statute, regulation, or rule of a State or political subdivision of a State that expressly requires comparable cybersecurity practices to protected covered critical infrastructure."[337] This section is followed by a savings clause that states: "Except as expressly provided in subsection (a) and section 105(e), nothing in this Act shall be construed to preempt the applicability of any other State law or requirement."[338]

Because the scope of "covered critical infrastructure" has yet to be determined, it is impossible to identify with specificity which state critical infrastructure laws would be preempted by this provision of the Cybersecurity Act of 2012. However, certain categories of state laws may be more likely to be preempted, such as those that directly regulate industrial facilities. For example, New Jersey has enacted the Toxic Catastrophe Prevention Act which was designed to prevent the release of hazardous substances from industrial plants and provide an abatement and evacuation plan in the event a catastrophic release occurs.[339] That act requires that an owner or operator of a covered facility establish a risk management program. Likewise, Maryland requires that any facility where hazardous materials are stored analyze the security of the facility every five years in accordance with rules adopted by the Department of State Police.[340] Similarly, New York requires the commissioner of the state division of homeland security to review security measures for all critical infrastructure relating to energy generation and transmission in the state every five years.[341] The state public service commission has the discretion whether to require the owners of these facilities to implement these plans. The application of these and other similar state requirements to covered critical infrastructure may be preempted if they are "comparable" to the risk-based performance standards to be established by the Secretary of DHS under the Cybersecurity Act of 2012.

It could also be argued that the broad language "comparable security practices," coupled with a savings clause that fails to carve out exceptions for state regulation of covered critical infrastructure, evidences Congress's intent to cover the whole field of protecting critical infrastructures.[342] Further, DHS may argue for a broad construction of this preemption provision as it has argued in the past that "the law of preemption recognizes that state laws must give way to Federal statutes and regulatory programs to ensure a unified and coherent national approach in areas where the Federal interests prevail—such as national security."[343] Because cybersecurity has been equated with national security, this deference theory could apply here.[344]

Cybersecurity legislation to encourage sharing of cybersecurity threat information may also preempt state laws. For example, all fifty states have included electronic communications in their respective wiretap laws which prohibit the interception and disclosure of certain communications.[345] However, S. 2102, the Cybersecurity Information Sharing Act contains an explicit preemption clause with respect to information sharing. Section 8(b) reads:

> This Act supersedes any law or requirement of a State or political subdivision of a State that restricts or otherwise expressly regulates the provision of cybersecurity services or the acquisition, interception, retention, use or disclosure of communications, records, or other information by private entities to the extent such law contains requirements inconsistent with this Act.[346]

The provisions of S. 2102, including this preemption clause, were also incorporated into Title VII of S. 2105. Similarly, H.R. 3523, the Cyber Intelligence Sharing and Protection Act (CISPA), contains a preemption clause providing that the bill would supersede "any statute of a State or political subdivision of a State that restricts or otherwise expressly regulates an activity authorized under subsection (b) [which authorizes private sector entities to use of cybersecurity systems to identify, obtain, or share cyber threat information]."[347] Likewise, H.R. 3674, the PRECISE Act, would "supersede[] any statute of a State or political subdivision of a State that restricts or otherwise expressly regulates the acquisition, interception, retention, use, or disclosure of communications, records, or other information" by private or government entities to the extent that such statute is inconsistent.[348] Similarly, the SECURE IT Act provides that it supersedes any state law that restricts or otherwise expressly regulates an activity authorized by the bill.[349]All of these provisions would likely be read to preempt the body of state wiretapping laws, to the extent that application of those laws would prevent the sharing of cybersecurity information authorized under the proposed legislation.[350] These bills could also preempt state data breach notification laws.[351] These bills could also preempt common law torts such as invasion of privacy or statutory remedies such as California's "Shine the Light" law, which regulates when and how a business can share a customer's personal information.[352]

State open records laws are another category that would likely be preempted under recent cybersecurity legislation being considered by Congress. Currently, states take a varied approach to exempting security information from state FOIA requirements.[353] Some states, including Indiana[354] and Alabama,[355] provide for specific disclosure exemptions for certain categories of information such as vulnerable assets or security plans. Others states, including Maryland, simply provide that anything protected under the federal FOIA statute is protected under their state statute.356 Still others have more broadly stated FOIA protections such as "in the public interest," as used in Arkansas.357 However, the PRECISE Act and the SECURE IT Act explicitly provide that cybersecurity information shared with state and local governments shall not be subject to any state or local law requiring disclosure of information or records.358 Both the Cybersecurity Act of 2012 and CISA similarly provide that information shared with a cybersecurity exchange designated under the bills would be exempt from FOIA "or any comparable State law."[359]

End Notes

[1] Critical infrastructure commonly refers to those entities that are so vital that their incapacitation or destruction would have a debilitating impact on national security, economic security, or the public health and safety. 42 U.S.C. §5195c(e). For more information, see CRS Report RL30153, Critical Infrastructures: Background, Policy, and Implementation, by John D. Moteff.

[2] See, e.g., Siobhan Gorman, Cybersecurity Bills Duel Over Rules for Firms, WALL ST. J., March 9, 2012, at A6.

[3] See also GOVERNMENT ACCOUNTABILITY OFFICE, Information Technology: Federal Laws, Regulations, and Mandatory Standards for Securing Private Sector Information Technology Systems and Data in Critical Infrastructure Sectors, GAO-08-1075R, September 16, 2008, available at http://www.gao.gov/assets/100/95747.pdf.

[4] The existing regulatory frameworks discussed here do not constitute an exhaustive list of all regulations applicable to critical infrastructure, but are only intended to provide some context for the following discussions.

[5] P.L. 109-295, §550 (codified at 6 U.S.C. §121 note). For a more detailed discussion of CFATS, see CRS Report R41642, Chemical Facility Security: Issues and Options for the 112th Congress, by Dana A. Shea.

[6] For a more detailed discussion of cybersecurity and electrical utilities, see CRS Report R41886, The Smart Grid and Cybersecurity—Regulatory Policy and Issues, by Richard J. Campbell.

[7] 46 U.S.C. §§70102-70103.

[8] For a more detailed discussion of critical infrastructure policy arguments, see CRS Report RL30153, Critical Infrastructures: Background, Policy, and Implementation, by John D. Moteff.

[9] E.g., Securing America's Future: The Cybersecurity Act of 2012 Before the S. Comm. on Homeland Security and Governmental Affairs, 112th Cong. (statement of former DHS Secretary Tom Ridge on behalf of U.S. Chamber of Commerce) ("New compliance mandates would drive up costs and misallocate business resources without necessarily increasing security.")

[10] 5 U.S.C. §701 et seq., see e.g., Nat'l Propane Gas Ass'n v. DHS, 534 F. Supp. 2d 16 (D.D.C. 2008) (denying temporary restraining order in action brought under APA claim for review of agency's designation of propane as chemical of interest for purposes of CFATS).

[11] For more information on judicial review of agency actions, see CRS Report R41546, A Brief Overview of Rulemaking and Judicial Review, by Vanessa K. Burrows and Todd Garvey.

[12] Gutierrez De Martinez v. Lamagno, 515 U.S. 417, 424 (1995) (quoting Bowen v. Michigan Academy of Family Physicians, 476 U.S. 667, 670 (1986)); see also McNary v. Haitian Refugee Center, Inc., 498 U.S. 479, 496 (1991); Abbott Laboratories v. Gardner, 387 U.S. 136 (1967); Citizens to Protect Overton Park v. Volpe, 401 U.S. 402 (1971); 28 U.S.C. §1331; but see Block v. Community Nutrition Institute, 467 U.S. 340, 349 (1984) (noting that "[t]he presumption favoring judicial review of administrative action ... may be overcome by specific language or specific legislative history that is a reliable indicator of congressional intent"). "The congressional intent necessary to overcome the presumption may also be inferred from contemporaneous judicial construction barring review and the congressional acquiescence in it ... or from the collective import of legislative and judicial history behind a particular statute," or from "inferences of intent drawn from the statutory scheme as a whole." Id.

[13] 5 U.S.C. §§702, 704.

[14] 5 U.S.C. §701.

[15] See Webster v. Doe, 486 U.S. 592 (1988); Oestereich v. Selective Service System, 393 U.S. 233 (1968).

[16] 5 U.S.C. §701(a)(1).

[17] 5 U.S.C. §701(a)(2).

[18] Citizens of Overton Park v. Volpe, 401 U.S. 402 (1971).

[19] Webster v. Doe, 486 U.S. 592 (1988).

[20] 50 U.S.C. §403-4a(e)(1).

[21] Webster, 486 U.S. at 600.

[22] 5 U.S.C. §706(2).

[23] Id. at §706(2)(E).

[24] Id. at §706(2)(A).

[25] See, e.g., Frontier Fishing Corp. v. Evans, 429 F. Supp. 2d 316, n.7 (citing Indus. Union Dep't v. API, 448 U.S. 607, 705 (1980) (Marshall, J., dissenting) (asserting that substantial evidence is more stringent, but is ultimately a deferential standard)).

[26] Consolidated Edison Co. v. NLRB, 305 U.S. 197, 229 (1938); NLRB v. Columbian Enameling & Stamping Co., 306 U.S. 292, 300 (1939)

[27] E.g., Kosmynka v. Polaris Industries, Inc., 462 F.3d 74, 79-82 (2d Cir. 2006) (upholding jury's finding that a manufacturer was negligent for failing to warn that its all-terrain vehicle might upend itself despite uncontested evidence that the manufacturer had received no reports of such incidents).

[28] Chevron v. Nat'l Resources Def. Council, 467 U.S. 837, 842-45 (1984).

[29] Id.

[30] A second issue with respect to enforcement is whether penalties would be limited to fines and other monetary penalties or whether injunctive relief may also be sought to compel

compliance or to stop a noncompliant facility from operating. For example, violations of CFATS can be punished by civil monetary penalties or an injunction to cease operations. 6 C.F.R. §27.300. Similarly, under MTSA, covered vessels and facilities without an approved security plan may be prohibited from operating. 46 U.S.C. §70103(c)(5). Questions may also arise regarding the types of investigative authorities that would be provided to the agency tasked with administering the regulatory scheme.

[31] Reese v. Philadelphia & R. R. Co., 239 U.S. 463, 465 (1915) ("The rule is well settled that a railroad company is not to be held as guaranteeing or warranting absolute safety to its employees under all circumstances, but is bound to exercise the care which the exigency reasonably demands in furnishing proper roadbed, tracks, and other structures. A failure to exercise such care constitutes negligence.").

[32] See RESTATEMENT (SECOND) OF TORTS §285 ("The standard of conduct of a reasonable man may be ... adopted by the court from a legislative enactment or an administrative regulation which does not so provide ...").

[33] See, e.g., Makas v. Hillhaven, Inc., 589 F. Supp. 736, 741 (M.D.N.C. 1984) ("Negligence per se in effect is a presumption that one who has violated a safety statute has violated its legal duty to exercise due care.").

[34] See, e.g., Resser v. Boise-Cascade Corp., 587 P.2d 80, 84 (Or. 1978) (violation of state law establishing speed limits at railroad crossing raises a rebuttable presumption of negligence).

[35] See, e.g., Burmaster v. Gravity Drainage Dist. No. 2, 448 So. 2d 162, 164 (La. Ct. App. 1984) (Occupational Safety and Health Act regulations and standards published by industry groups warrant consideration as evidence of standard of care, even if they are not controlling).

[36] See CRS Report R41406, The Freedom of Information Act and Nondisclosure Provisions in Other Federal Laws , by Gina Stevens and CRS Report RL33670, Protection of Security-Related Information, by Gina Stevens and Todd B. Tatelman.

[37] 5 U.S.C. §552.

[38] National Freedom of Information Coalition, State Freedom of Information Laws (2012), at http://www.nfoic.org/ state-freedom-of-information-laws.

[39] As an example, in Milner v. Dept. of the Navy, 131 S. Ct. 1259 (2011), the Supreme Court limited the scope of FOIA Exemption 2 (the Court held that "Exemption 2, consistent with the plain meaning of the term "personnel rules and practices," encompasses only records relating to issues of employee relations and human resources."). Id. at 1271. See U.S. Dep't of Justice, Exemption 2 After the Supreme Court's Ruling in Milner v. Department of the Navy, at http://www.justice.gov/oip/foiapost/2011foiapost15.html.

[40] Testimony of David Sobel, Electronic Privacy Information Clearinghouse before the U.S. Congress, House Committee on Energy and Commerce, Subcommittee on Oversight and Investigations, Creating The Department of Homeland Security: Consideration of the Administration's Proposal, 107th Cong., 2nd sess., June 25 and July 9, 2002, Serial No. 107-113 (Washington: GPO, 2002), p. 258.

[41] 5 U.S.C. §552.

[42] 5 U.S.C.§552(a)(1).

[43] 5 U.S.C. §552(a)(2).

[44] Excluded from the act's coverage are special categories of law enforcement records related to criminal law investigations or proceedings, informant records, and records maintained by the FBI pertaining to foreign intelligence, counterintelligence or international terrorism. 5 U.S.C. §552(c)(1), (c)(2), (c)(3).

[45] See Dep't of the Air Force v. Rose, 425 U.S. 352, 361 (1976) (holding that "limited exemptions do not obscure the basic policy that disclosure, not secrecy, is the dominant objective of the Act").

[46] 5 U.S.C. §552(b)(3).

[47] 5 U.S.C. §552(b)(3).

[48] U.S. Department of Justice, Statutes Found to Qualify under Exemption 3 of the FOIA, (August 2011), available at http://www.justice.gov/oip/exemption3.pdf.

[49] Examples of FOIA exemption 3 statutes are the Aviation and Transportation Security Act of 2001 (ATSA) which designates 16 categories of sensitive security information and includes information submitted pursuant to a requirement and information voluntarily submitted, P.L. 107-71, codified at 49 U.S.C. §40119; the Critical Infrastructure Information Act of 2002 (CIIA) which provides confidentiality protections for critical infrastructure information voluntarily submitted to DHS, P.L. 107-296, codified at 6 U.S.C. §133 et seq.; the Maritime Transportation Security Act of 2002 (MTSA) which requires covered entities to submit information to the federal government, P.L. 107-295; and the Safe Drinking Water Act (SDWA) , as amended, which requires community water systems to perform vulnerability analyses of their facilities and includes protections for vulnerability assessments. P.L. 107-188, 42 U.S.C. §300i-2.

[50] Department of Justice, "Homeland Security Law Contains New Exemption 3 Statute," FOIA Post (2003).

[51] 6 C.F.R. §29.2(b).

[52] See U.S. Dept. of Homeland Security, Protected Critical Infrastructure Information (PCII) Program, at http://www.dhs.gov/files/programs/editorial_0404.shtm; PCII Program and Procedures Guidance Manual (April 2009) at http://www.dhs.gov/xlibrary/assets/pcii_program_procedures_manual.pdf.

[53] See Fed. R. Evid. 501.

[54] P.L. 112-8, §1091, 125 Stat. 1604.

[55] 5 U.S.C. §556(e).

[56] Id.

[57] 5 U.S.C. §557(d)(1). For example, under CFATS, during an adjudication ex-parte communications between the department and the chemical facility is not permitted. 6 C.F.R. §27.320.

[58] 5 U.S.C. §557(d)(1)(E).

[59] 5 U.S.C. §557(d)(1)(C).

[60] H.R. 3674, §2 (new §227 of the Homeland Security Act of 2002 (HSA)).

[61] Id. (new HSA §227(f)).

[62] Id. (new HSA §227(g)).

[63] S. 2105, §102.

[64] S. 2105, §103.

[65] S. 2105, §104.

[66] S. 2105, §104(d).

[67] Owners of critical infrastructure can also self-designate or request that their systems or assets be considered covered critical infrastructure. S. 2105, §103(b)(4).

[68] S. 2105, §103(b)(1)(C).

[69] S. 2105, §103(b)(1)(C)(ii).

[70] That term is defined in the bill to mean "a commercial item that organizes or communicates information electronically." S. 2105, §2(1).

[71] S. 2105, §103(b)(2).

[72] S. 2105, §103(c). Many other Secretarial decisions, such as the determination that an existing regulatory scheme is inadequate, would appear to have sufficient judicially manageable standards to qualify for judicial review under the APA. But see S. 2105, §104(f)(1), authorizing the President to exempt any covered critical infrastructure from performance standards, if the President determines that a sector specific regulatory agency has sufficient requirements to protect against the identified risks. Such a decision may not be subject to judicial review because the APA does not generally apply to decisions made by the President. Franklin v. Massachusetts, 505 U.S. 788, 800 (1992).

[73] S. 2105, §103(c).

[74] S. 2105, §105(c)(1)(B).

[75] S. 2105, §105(c)(1)(A). Companies that can demonstrate that their covered critical infrastructure are sufficiently secured against the risks identified would only have to certify every three years. S. 2105, §105(c)(4).

[76] S. 2105, §105(d)(2).

[77] S. 2105, §105(e).

[78] Id.

[79] S. 2105, §§101(b), 105(d)(3)(A), 107(a)(1).

[80] S. 2105, §107(b). The CIIA consists of a group of provisions that address the circumstances under which the Department of Homeland Security may obtain, use, and disclose critical infrastructure information as part of a critical infrastructure protection program. It was enacted, in part, to respond to the need for the federal government and owners and operators of the nation's critical infrastructures to share information on vulnerabilities and threats, and to promote information sharing between the private and public sectors in order to protect critical assets.

[81] S. 2105, §107(b).

[82] S. 2105, §107(a)(2).

[83] S. 2105 §301 (new HSA §242(a)).

[84] Id. (new HSA §242(d)).

[85] Id. (new HSA §242(c)).

[86] See Id. (new HSA §241(12)). Intelligence community has the meaning given in 50 U.S.C. §401a(4), which includes the Office of Director of National Intelligence; the Central Intelligence Agency; the National Security Agency; certain elements of the Department of Defense; intelligence elements of the military branches, the Federal Bureau of Investigation, the Drug Enforcement Agency, and the Departments of Energy, State, Homeland Security, and Treasury.

[87] S. 2105 §301 (new HSA §242(g)(2)).

[88] S. 2105 §§102(a)(2), and 301 (new HSA §242(e)(2)).

[89] S. 2105 §301 (new HSA §242(e)(6)(B)).

[90] S. 2105 §105(b).

[91] S. 2105 §301 (new HSA §242(e)(8)).

[92] S. 2105 §109.

[93] S. 2105 §104(f).

[94] Title III defines covered critical infrastructure as "as system or asset designated by the Secretary ... in accordance with the procedure established under section 103 of the Cybersecurity Act of 2012." S. 2105 §301 (new HSA §241(3)). This definition makes no mention of the exemption process that takes place after the designation is determined, as laid out in Section 104. See S. 2105 §104(f).

[95] S. 2105 §301 (new HSA §243(c)(1)(B)).

[96] This section focuses on the constitutional concerns with EINSTEIN under the Fourth Amendment. Although statutes such as the Electronic Communications Privacy Act of 1986, P.L. 99-508, 100 Stat. 1848, and the Privacy Act of 1974, 5 U.S.C. §522a, may be implicated, they will not be discussed here.

[97] DEP'T OF HOMELAND SECURITY, PRIVACY IMPACT ASSESSMENT: EINSTEIN PROGRAM, at 3 (2004) (hereinafter EINSTEIN 1 PRIVACY IMPACT ASSESSMENT), available at http://www.dhs.gov/xlibrary/assets/privacy/privacy_pia_eisntein.pdf.

[98] Id. at 1.

[99] Office of Management and Budget, Memorandum for the Heads of Executive Departments and Agencies: Implementation of Trusted Internet Connections (TIC) (November 20, 2007), available at http://www.whitehouse.gov/ sites/default/files/omb/assets/omb/ memoranda/ fy2008/m08-05.pdf.

[100] EINSTEIN 1 PRIVACY IMPACT ASSESSMENT, supra note 94 at 4.

[101] Id. at 6-7. An IP address is a unique identifier used by most computers when sending data over the Internet. It is akin to a personal telephone number or street address. See Stephanie Crawford, What is an IP address?, HOW STUFF WORKS, http://computer. howstuffworks. com/internet/basics/question549.htm.

[102] See http://www.us-cert.gov/cas/techalerts/ for an example of cybersecurity alerts provided to the public.

[103] DEP'T OF HOMELAND SECURITY, PRIVACY IMPACT ASSESSMENT: EINSTEIN 2, at 1 (2008) (hereinafter EINSTEIN 2 PRIVACY IMPACT ASSESSMENT), available at http://www.dhs.gov/xlibrary/assets/privacy/privacy_pia_einstein2.pdf.

[104] Id. at 9. For more information on intrusion detection systems, see NAT'L INSTITUTE OF STANDARDS AND TECH., GUIDE TO INTRUSION DETECTION AND PREVENTION SYSTEMS (IDPS) (2007) (Pub. No. 800-94), available at http://csrc.nist. gov/publications/nistpubs/800-94/SP800-94.pdf (hereinafter "NIST REPORT").

[105] NIST REPORT, supra note 101, at 9-5.

[106] EINSTEIN 2 PRIVACY IMPACT ASSESSMENT, supra note 100, at 10.

[107] According to DHS, the name of the agency is classified. DEP'T OF HOMELAND SECURITY, PRIVACY IMPACT ASSESSMENT: INITIATIVE THREE EXERCISE, at 3 (2010) (hereinafter EINSTEIN 3 PRIVACY IMPACT ASSESSMENT) available at http://www.dhs.gov/xlibrary/assets/privacy/privacy_pia_nppd_initiative3.pdf.

[108] Id. at 3.

[109] Id. at 4.

[110] See, e.g., DEP'T OF HOMELAND SECURITY, PRIVACY COMPLIANCE REVIEW OF THE EINSTEIN PROGRAM (2012) (hereinafter EINSTEIN PRIVACY COMPLIANCE REVIEW), available at http://www.dhs.gov/xlibrary/assets/privacy/ privacy_privcomrev_ nppd_ein.pdf; THE CONSTITUTION PROJECT, RECOMMENDATIONS FOR THE IMPLEMENTATION OF A COMPREHENSIVE AND CONSTITUTIONAL CYBERSECURITY POLICY (2012) (hereinafter THE CONSTITUTION PROJECT), available at http://www.constitutionproject. org/pdf/TCPCybersecurityReport.pdf; Jack Goldsmith, The Cyberthreat, Government Network Operations, and the Fourth Amendment (2010), available at http://www.brookings.edu/papers/ 2010/1208_4th_ amendment_ goldsmith.aspx.

[111] U.S. CONST. amend. IV.

[112] Camara v. Mun. Ct., 387 U.S. 523, 528 (1967).

[113] Kyllo v. United States, 533 U.S. 27, 32-33 (2001).

[114] This formulation for determining whether a search of seizure occurred derives from Justice Harlan's concurrence in Katz v. United States, 389 U.S. 347, 361 (1967) (Harlan, J., concurring).

[115] Texas v. Brown, 460 U.S. 730, 739 (1983).

[116] Mincey v. United States, 437 U.S. 385, 390 (1978). Probable cause has been defined as "the facts and circumstances within the officers' knowledge and of which they had reasonably trustworthy information are sufficient in themselves to warrant a man of reasonable caution in the belief that an offense has been or is being committed." Brinegar v. United States, 338 U.S. 160, 175 (1948).

[117] United States v. Forrester, 512 F.3d 500, 511 (9th Cir. 2007) (holding no reasonable expectation of privacy in the to/from line addresses of e-mails and IP address of websites visited); United States v. Christie, 624 F.3d 558, 574 (3rd Cir.) (holding no reasonable expectation of privacy in IP address); United States v. Perrine, 518 F.3d 1196, 1205 (10th Cir.) (holding no reasonable expectation of privacy in Internet subscriber information given to Internet service provider).

[118] United States v. Forrester, 512 F.3d at 511. A pen register is a device that records the numbers dialed from a telephone. 18 U.S.C. §3127(3).

[119] Id. at 510. In Smith v. Maryland, the Court held that the use of a pen register—a device that obtains the telephone numbers dialed from a certain phone—was not a search under the Fourth Amendment. 442 U.S. 735, 745-46 (1979).

[120] Forrester, 512 F.3d at 510.

[121] United States v. Warshak, 631 F.3d 266, 287 (6th Cir. 2010) (internal quotation marks omitted).

[122] United States v. Lifshitz, 369 F.3d 173, 190 (2d Cir. 2004).

[123] City of Ontario v. Quon, 130 S. Ct. 2619, 2630 (2010).

[124] See Orin Kerr, Applying the Fourth Amendment to the Internet: A General Approach, 62 STAN. L. REV. 1005, 1022- 29) (2010) (analogizing the content/non-content distinction developed in the Fourth Amendment letter and telephone cases with Internet communications).

[125] Ex parte Jackson, 96 U.S. 727, 733 (1878); Forrester, 512 F.3d at 511 (citing Ex parte Jackson, 96 U.S. at 733).

[126] Ex parte Jackson, 96 U.S. at 733. The constitutional guaranty of the right of the people to be secure in their papers against unreasonable searches and seizures extends to their papers, thus closed against inspection, wherever they may be. Whilst in the mail, they can only be opened and examined under like warrant, issued upon similar oath or affirmation, particularly describing the thing to be seized, as is required when papers are subjected to search in one's own household. No law of Congress can place in the hands of officials connected with the postal service any authority to invade the secrecy of letters and such sealed packages in the mail; and all regulations adopted as to mail matter of this kind must be in subordination to the great principle embodied in the fourth amendment of the Constitution.

Id.

[127] Kerr, supra note 121, at 1023-24.

[128] Katz v. United States, 389 U.S. 347, 359 (1967)

[129] Smith v. Maryland, 442 U.S. 735, 745-46 (1979).

[130] EINSTEIN PRIVACY COMPLIANCE REVIEW, supra note 107, at 5.

[131] See Legal Issues Relating to the Testing, Use, and Deployment of an Intrusion-Detection System (EINSTEIN 2.0) to Protect Unclassified Computer Networks in the Executive

Branch, 33 Op. O.L.C. 1, *11 (2009) (hereinafter Legal Issues Relating to EINSTEIN 2.0), available at http://www.justice.gov/olc/2009/e2-issues.pdf.

[132] There is also a third category of cases: where a federal employee sends a communication while on the federal network to a private person. Because the principles that apply to communications from a private person to a federal employee are the same as the principles that apply to communications from a federal employee to a private person, these two categories will be discussed jointly.

[133] City of Ontario v. Quon, 130 S. Ct. 2619, 2624 (2010). For an in-depth treatment of Quon, see CRS Report R41344, Public Employees' Right to Privacy in Their Electronic Communications: City of Ontario v. Quon in the Supreme Court , by Charles Doyle.

[134] Quon, 130 S. Ct. at 2625.

[135] Id. at 2630.

[136] Id. at 2630.

[137] Id. at 2631.

[138] Id. (internal citations omitted).

[139] Legal Issues Relating to EINSTEIN 2.0, supra note 128, at *11.

[140] Id. at 32-33.

[141] Kerr, supra note 121, at 1031.

[142] See also THE CONSTITUTION PROJECT, supra at note 107, at 14 ("For federal employees, the analysis that employees consent to having Einstein monitor communications is likely reasonable given the overwhelming importance of protecting key federal agency networks.").

[143] Quon, 130 S. Ct. at 2631.

[144] Id.

[145] Another possible approach is that taken in United States v. Jones, 565 U.S. ___ (2012) (slip op.), in which the Court held that a physical intrusion into a constitutionally protected area—there, the defendant's car (an effect)—coupled with an attempt to obtain information, was a Fourth Amendment search. If a court concluded that an e-mail is a paper (or packet of data, an effect), protected under the Fourth Amendment's catalog of protected areas (persons, houses, papers, and effects), the Jones physical intrusion analysis may call into question whether EINSTEIN's surveillance is constitutionally permissible.

[146] Legal Issues Relating to EINSTEIN 2.0, supra note 128, at 35-36 (citing Smith v. Maryland, 442 U.S. 735, 743-44) (1979).

[147] United States v. Miller, 425 U.S. 435 (1976) holding that financial statements and deposit slips transmitted to bank were not protected from police inquiry because they had been turned over to a third party); Smith, 442 U.S. 735. It should be noted that in United States v. Jones, Justice Sotomayor opined that it "may be necessary to reconsider the premise that an individual has no reasonable expectation of privacy in information voluntarily disclosed to third parties." United States v. Jones, 565 U.S. ___, 5 (Sotomayor, J., concurring in the judgment and the opinion).

[148] United States v. White, 401 U.S. 745, 750 (1971) (holding that the Fourth Amendment "affords no protection to a wrongdoer's misplaced belief that a person to whom he voluntarily confides his wrongdoing will not reveal it.") (internal quotation marks omitted).

[149] Legal Issues Relating to EINSTEIN 2.0, supra note 128, at 36-37.

[150] Smith, 442 U.S. at 741 (quoting United States v. N.Y. Tel. Co., 434 U.S. 159, 167 (1977)).

[151] Id. at 742.

[152] United States v. Forrester, 512 F.3d 500, 511 (9th Cir. 2007).

[153] United States v. Warshak, 631 F.3d 266, 285 (6th Cir. 2007).

[154] United States v. U.S. Dist. Ct., 407 U.S. 297, 313 (1972).
[155] Legal Issues Relating to EINSTEIN 2.0, supra note 128, at 39.
[156] SEC v. Jerry T. O'Brien, Inc., 467 U.S. 735, 743 (1984).
[157] Chandler v. Miller, 520 U.S. 305, 308 (1997).
[158] Nat'l Treasury Emplys. Union v. Von Raab, 489 U.S. 656, 665-66 (1989).
[159] Id.
[160] Michigan Dep't of State Police v. Sitz, 496 U.S. 444, 455 (1990).
[161] United States v. Ramsey, 431 U.S. 606 (1977).
[162] Illinois v. Lidster, 540 U.S. 419, 428 (2004).
[163] City of Indianapolis v. Edmond, 531 U.S. 32, 38 (2000).
[164] Legal Issues Relating to EINSTEIN 2.0, supra note 128, at 54.
[165] Goldsmith, supra note 107, at 14.
[166] THE CONSTITUTION PROJECT, supra note 107, at 16.
[167] Goldsmith, supra note 107, at 13.
[168] Electronic Communications Privacy Act of 1986, P.L. 99-508, 100 Stat. 1848.
[169] Privacy Act of 1974, P.L. 93-579, 88 Stat. 1896.
[170] EINSTEIN 1 PRIVACY IMPACT ASSESSMENT, supra note 94, at 9.
[171] EINSTEIN 2 PRIVACY IMPACT ASSESSMENT, supra note 100, at 12.
[172] Id.
[173] EINSTEIN PRIVACY COMPLIANCE REVIEW, supra note 107, at 4.
[174] Id. at 7.
[175] The Foreign Intelligence Surveillance Court is a comprised of 11 federal district court judges who are designated by the Chief Justice to hear applications for surveillance orders authorized under the Foreign Intelligence Surveillance Act of 1978. 50 U.S.C. §1803.
[176] Goldsmith, supra note 107, at 14.
[177] THE CONSTITUTION PROJECT, supra note 107, at 28.
[178] Id.
[179] S. 2105, 112th Cong. §3553 (2012).
[180] S. 2105 §301 (new HSA §242(a)).
[181] "Federal information infrastructure" is defined as "information and information systems that are owned, operated, controlled, or licensed by, or on behalf of, a Federal agency ... " but explicitly excludes systems used by the Department of Defense, the military, and an element of the intelligence community. S. 2105 §301 (new HSA §241(8)).
[182] Id. (new HSA §241(8)(B), 242(e)(3)).
[183] Id. (new HSA §243(b)(1)(B), (c)(1)(B)).
[184] Id. (new HSA §243(c)(1)(C)).
[185] Id. (new HSA §242(e)(3)).
[186] Id. (new HSA §242(e)(2)).
[187] Id. §201 (new 44 U.S.C. §3553(b)(3), (f)).
[188] See Id. (new 44 U.S.C. §3553(f)).
[189] See S. 2105 §301 (new HSA §241(8)) (defining "Federal information infrastructure").
[190] Id. (new HSA §243(b)(2)(B)-(C)).
[191] H.R. 3674, 112th Cong. (2012) (new HSA §226(c)(1).
[192] See EINSTEIN Overview, supra p. 14.
[193] The term "countermeasure" means "automated actions with defensive intent to modify or block data packets associated with electronic or wire communications, Internet traffic, program code, or other system traffic transiting to or from or stored on an information

system for the purpose of protecting the information system from cybersecurity threats." H.R. 3674 (new HSA §226(g)(1).

[194] Id. (new HSA §226(c)(1)(A)).

[195] Legal Issues Relating to EINSTEIN 2.0, supra note 128, at 14-15.

[196] See Department of Homeland Security, Web Site Privacy Policy, http://www.dhs.gov/xutil/gc_1157139158971.shtmm (last visited April 19, 2012).

[197] See United States v. Miller, 425 U.S. 435 (1976).

[198] Smith v. Maryland, 442 U.S. 735, 741 (1979).

[199] H.R. 3674 (new HSA §226(c)(1)(B)).

[200] H.R. 4257, 112th Cong. (2012) (new 44 U.S.C. §3554).

[201] Id. (new 44 U.S.C. §3553).

[202] Id. (new 44 U.S.C. §3554).

[203] Id. (new 44 U.S.C.§3555).

[204] See CRS Report 98-326, Privacy: An Overview of Federal Statutes Governing Wiretapping and Electronic Eavesdropping, by Gina Stevens and Charles Doyle for a more detailed discussion of the federal laws governing wiretapping and electronic eavesdropping, along with appendices including copies of the texts of ECPA and FISA. See also CRS Report R41733, Privacy: An Overview of the Electronic Communications Privacy Act, by Charles Doyle.

[205] 18 U.S.C. §§2510-2522.

[206] 18 U.S.C. §§2701-2712.

[207] 18 U.S.C. §§3121-3127. Pen registers capture the numbers dialed on a telephone line; trap and trace devices identify the originating number of a call on a particular phone line. See 18 U.S.C. §3127(3)-(4). The USA PATRIOT Act enlarged the coverage of the Pen Register Statute to include sender/addressee information relating to email and other forms of electronic communications. P.L. 107-56, §216(c)(2).

[208] 18 U.S.C. §2511.

[209] 18 U.S.C. §2511(2)(a)(i), (h).

[210] 18 U.S.C. §2511(2)(a)(ii).

[211] 18 U.S.C. §2511(2)(b).

[212] 18 U.S.C. §2511(2)(c).

[213] 18 U.S.C. §2702(a)(1).

[214] 18 U.S.C. §2702(a)(2).

[215] 18 U.S.C. §2702(a)(3).

[216] 18 U.S.C. §2702(b) (emphasis added). The record disclosure exceptions are similar. 18 U.S.C. 2702(c).

[217] 18 U.S.C. §3121 ("Except as provided in this section, no person may install or use a pen register or a trap and trace device without first obtaining a court order under section 3123 of this title or under the Foreign Intelligence Surveillance Act of 1978 (50 U.S.C. 1801 et seq.)").

[218] 18 U.S.C. §3121(b).

[219] 18 U.S.C. §3125(a).

[220] See, e.g., Aaron J. Burstein, Amending the ECPA to Enable a Culture of Cybersecurity Research, 22 HARV. J.L. & TECH. 167 (2008).

[221] 18 U.S.C. §2511(2)(a)(i).

[222] 15 U.S.C. §§1, 4, 15, 26.

[223] 15 U.S.C. §1.

[224] Standard Oil Co. of N.J. v. U.S., 221 U.S. 1, 60 (1911) (interpreting the language of Section One to require that in order for restraints in trade to be considered unlawful, the methods used to restrain the market must be undue or unreasonable).

[225] See Fed. Trade Comm'n & U.S. Dep't of Justice, Antitrust Guidelines for Collaborations Among Competitors (2000), available at http://www.ftc.gov/os/2000/04/ ftcdojguidelines. pdf (hereinafter Competitor Collaboration Guidelines).

[226] Nat'l Soc'y of Prof'l Eng'rs v. United States, 435 U.S. 679, 692 (1978).

[227] Business Electronics Corp. v. Sharp Electronics Corp., 485 U.S. 717, 723 (1988).

[228] Competitor Collaboration Guidelines, supra note 208, at 3.

[229] Id.

[230] Continental T.V. Corp. v. GTE Sylvania Corp., 433 U.S. 36, 49 (1977).

[231] Competitor Collaboration Guidelines, supra note 208, at 6-7.

[232] Id. at 1.

[233] Id.

[234] Id. at 12.

[235] See Letter from Joel I. Klein, Assistant Attorney General, Department of Justice, Antitrust Division, to Barbara Greenspan, Associate General Counsel, Electric Power Research Institute, Inc. (October 2, 2000) available at http://justice.gov/atr/public/busreview/ 6614.htm.

[236] 28 C.F.R. §50.6.

[237] Letter from Joel I. Klein, Assistant Attorney General, Department of Justice, Antitrust Division, to Barbara Greenspan, Associate General Counsel, Electric Power Research Institute, Inc. (October 2, 2000) available at http://justice.gov/atr/public/ busreview/ 6614.htm.

[238] 15 U.S.C. §§1, 4; 18 U.S.C. §§2511, 2701, 3121.

[239] 15 U.S.C. §§15, 26; 18 U.S.C. §§2520, 2707.

[240] See, e.g., First Electric Cooperative Corp. v. Pinson, 642 S.W.2d 301, 303 (Ark. 1982) ("there is no negligence in not guarding against a danger which there is no reason to anticipate").

[241] Id.

[242] E.g., Schuldies v. Service Machine Co., 448 F. Supp. 1196, 1199 (E.D. Wis. 1978) ("a person fails to exercise ordinary care when, without intending to do any wrong, he does an act or omits a precaution under circumstances in which a person of ordinary intelligence and prudence ought reasonably to foresee that such act or omission will subject the interests of another to an unreasonable risk of harm").

[243] E.g., Rodriguez v. New Haven, 439 A.2d 421, 424 (Conn. 1981) ("knowledge of a dangerous condition generally requires greater care to meet the standard of reasonable care").

[244] See CRS Report R41406, The Freedom of Information Act and Nondisclosure Provisions in Other Federal Laws , by Gina Stevens and CRS Report RL33670, Protection of Security-Related Information, by Gina Stevens and Todd B. Tatelman.

[245] 5 U.S.C. §552.

[246] See Fed. R. Evid. 501.

[247] See e.g., Center for Democracy & Technology, Concerns Mount Over Unresolved Privacy Issues in CISPA, https://www.cdt.org/blogs/greg-nojeim/1804concerns-mount-over-unre solved-privacy-issues-cispa.

[248] H.R. 3523, §2; H.R. 4263, §102(1).

[249] Cybersecurity Information Sharing and the Freedom of Information Act: Hearing Before the S. Comm. on the Judiciary, 112th Cong. (2012) (statement of Paul Rosenzweig, Visiting

Fellow, The Heritage Foundation), available at http://www.judiciary.senate.gov/pdf/12-3-13RosenzweigTestimony.pdf.

[250] Under the Posse Comitatus Act, the military is prohibited from executing domestic laws. 18 U.S.C. §1385.

[251] Michelle Richardson, Cybersecurity Information Sharing Legislation and Privacy Implications in the 112th Congress, AMERICAN CIVIL LIBERTIES UNION (April 16, 2012), http://www.aclu.org/files/assets/ aclu_interested_persons_memo__re__ cyber_leg_info_sharing_april_16_2012.pdf.

[252] Greg Nojeim, Cybersecurity's 7-Step Plan for Internet Freedom, CENTER FOR DEMOCRACY AND TECHNOLOGY (March 28, 2012), https://www.cdt.org/blogs/greg-nojeim/2803cybersecuritys-8-step-plan-internet-freedom.

[253] Id.

[254] Under 50 U.S.C. §401a(4), the IC is comprised of the following offices: the Office of the Director of National Intelligence; the Central Intelligence Agency; the National Security Agency; the Defense Intelligence Agency; the National Geospatial-Intelligence Agency; the National Reconnaissance Office; other offices within the Department of Defense for the collection of specialized national intelligence through reconnaissance programs; the intelligence elements of the Army, the Navy, the Air Force, the Marine Corps, the Coast Guard, the Federal Bureau of Investigation, the Drug Enforcement Administration, and the Department of Energy; the Bureau of Intelligence and Research of the Department of State; the Office of Intelligence and Analysis of the Department of the Treasury; the Office of Intelligence and Analysis of the Department of Homeland Security; and such other elements as may be designated by the President, or designated jointly by the Director of National Intelligence and the head of the department or agency concerned, as an element of the intelligence community.

[255] H. Permanent Select Comm. on Intelligence, Bipartisan Cybersecurity Bill Clears Key Hurdle, December 1, 2011, available at http://intelligence.house.gov/press-release/ bipartisan-cybersecurity-bill-clears-key-hurdle. See also Lauren Gardner, House Panel Backs Cybersecurity Bill with Stronger Privacy Provisions, CQ, December 1, 2011, available at http://www.cq.com/news.do.

[256] H.R. 3523 (April 16, 2012 - Discussion Draft), available at http://intelligence. house.gov/sites/ intelligence.house.gov/files/documents/04162012HR3523.pdf.

[257] U.S. Congress, House Permanent Select Committee on Intelligence, The Cyber Intelligence Sharing and Protection Act, report to accompany H.R. 3523, 112th Cong., 2nd sess., 2012, H.Rept. 112-445.

[258] Http://www.rules.house.gov/Media/file/PDF_112_2/LegislativeText/CPRT-112-HPRT-RU 00- HR3523.pdf.

[259] H.R. 3523, §2 (new §1104(a)(1) of the National Security Act of 1947 (NSA)).

[260] Id. (new NSA §1104(b)(f)(6)).

[261] Id. (new NSA §1104(a)(2)).

[262] Id. (new NSA §1104(a)(2)(C)).

[263] Id. (new NSA §1104(a)(3)).

[264] Id. (new NSA §§1104(b)(1), (2)).

[265] Id.

[266] "Congress sometimes seeks to underscore the primacy of a statutory directive by stating that it is to apply "notwithstanding" the provisions of another, specified statute or class of statutes. Courts take into account this expressed intent to override the provisions specified in a "notwithstanding" clause,.. but when the clause purports to override "any other provision of

law," its preclusive scope often is unclear....As a rule, though, it might be more effective to spell out which other laws are to be disregarded,... and it must be kept in mind, of course, that no "notwithstanding" clause can foreclose subsequent legislation that supersedes it expressly or implicitly." [citations omitted]. CRS Report 97-589, Statutory Interpretation: General Principles and Recent Trends, by Larry M. Eig.

[267] H.R. 3523, (new NSA §1104(b)(2)(A)). Minimization typically refers to limitations on what information is acquired; how it is acquired; how it is maintained; who has access to it within the capturing agency and under what circumstances; to whom and under what circumstances it is disclosed beyond the capturing agency; how long it is preserved; and when and under what circumstances it is expunged. For a discussion of minimization procedures in the context of national security letters, see CRS Report R41619, National Security Letters: Proposals in the 112th Congress, by Charles Doyle.

[268] Id. (new NSA §1104(b)(2)(B)).

[269] Id. (new NSA §1104(b)(2)(C)).

[270] Id. (new NSA §1104(b)(2)(C)(ii)).

[271] Id. (new NSA §1104(b)(3)(A)).

[272] Id. (new NSA §1104(c)(1)).

[273] Id. (new NSA §1104(c)(2)).

[274] Id. (new NSA §1104(c)(3)).

[275] H.R. 3674, available at http://homeland.house.gov/markup/markup-hr-3674-promoting-and-enhancingcybersecurity-and-information-sharing-effectiveness.

[276] S. 2102, §2; S. 2105, §701.

[277] S. 2102, §3; S. 2105, §702.

[278] S. 2102, 8(a)(5); S. 2105, §707(a)(5).

[279] S. 2102, §4; S. 2105, §703. The Secretary of DHS would also be required to designate a lead cybersecurity exchange to serve as the focal point within the federal government for cybersecurity information sharing. S. 2102, §4(c); S. 2105, §703(c).

[280] S. 2102, §5; S. 2105, §704. Classified threat information may only be shared with certified entities with adequate security clearances. Security clearances may be granted to certified entities and employees of certified entities. S. 2102, §6; S. 2105, §705.

[281] S. 2102, §§5(d), (e); S. 2105, §§704(d), (e).

[282] S. 2102, §5(f); S. 2105, §704(f).

[283] S. 2102, §7(c); S. 2105, §706(c).

[284] S. 2102, §5(g)(1)(A); S. 2105, §704(g)(1)(A).

[285] S. 2102, §5(g)(1)(B); S. 2105, §704(g)(1)(B).

[286] S. 2102, §5(g)(2); S. 2105, §704(g)(2).

[287] S. 2102, §5(g)(3); S. 2105, §704(g)(3).

[288] S. 2102, §5(g)(3)(B); S. 2105, §704(g)(3)(B).

[289] S. 2102, §5(g)(4); S. 2105, §704(g)(4).

[290] S. 2102, §7(a); S. 2105, §706(a).

[291] S. 2102, §7(b); S. 2105, §706(b).

[292] S. 2102, §7(f); S. 2105, §706(f).

[293] S. 2102, §7(e); S. 2105, §706(e).

[294] S. 2102, §7(g); S. 2105, §706(g).

[295] S. 2105 §703(a)-(b).

[296] S. 2105 §301 (new HSA §243).

[297] S. 2105 §301 (new HSA §243(b)(1)(B), (c)(1)(B)).

[298] S. 2105 §301 (new HSA §243(c)(1)(C)).

[299] S. 2105 §703(c)(3)(A). Until this designation is finalized, the National Cybersecurity and Communications Integration Center (NCCIC) would serve as the interim lead exchange. S. 2105 §703(c)(3)(B).
[300] S. 2105 §703(d).
[301] S. 2105 §704(g).
[302] S. 2105 §704(c).
[303] S. 2105 §704(b), (g).
[304] See 5 U.S.C. §552.
[305] S. 2105 §704(d)(1), (e).
[306] S. 2105 §706(a)(2).
[307] S. 2105 §705(a).
[308] S. 2105 §301 (new HAS §243(a)(2)).
[309] See S. 2105 §301 (new HSA §243(c)(1)(B).
[310] S. 2151, §102(a)(1).
[311] Id.
[312] The bill defines the term cybersecurity center to mean the Department of Defense Cyber Crime Center, the Intelligence Community Incident Response Center, the United States Cyber Command Joint Operations Center, the National Cyber Investigative Joint Task Force, the National Security Agency/Central Security Service Threat Operations Center, the National Cybersecurity and Communications Integration Center, and any successor center. S. 2151, §101(5).
[313] S. 2151, §102(a)(2).
[314] S. 2151, §102(a)(3).
[315] S. 2151, §§102(e)(1), (2).
[316] S. 2151, §102(e)(3).
[317] S. 2151, §102(b)(1).
[318] Id.
[319] S. 2151, §102(b)(2).
[320] S. 2151, §102(d)(1)(B).
[321] S. 2151, §102(c)(1)(A).
[322] S. 2151, §102(c)(1)(B).
[323] S. 2151, §102(c).
[324] S. 2151, §102(c)(2) (requiring prior consent for disclosures to state, tribal, and local governments for criminal prevention, investigation, or prosecution); S. 2151, §102(c)(3) (requiring prior consent for disclosures outside of the federal government); S. 2151, §102(c)(7) (requiring prior consent for subsequent disclosures of information shared with state, tribal, and local governments).
[325] S. 2151, §§102(c)(4), (5), (7).
[326] S. 2151, §102(c)(6).
[327] S. 2151, §102(c)(8). However, the procedures to implement the SECURE IT Act would not be considered "regulatory" for the purposes of this limitation. Id.
[328] S. 2151, §102(g)(1)(A).
[329] S. 2151, §102(g)(1)(B).
[330] S. 2151, §102(g)(2).
[331] Hillsborough County v. Automated Med. Labs., Inc., 471 U.S. 707, 713 (1985).
[332] Gade v. Nat'l Solid Wastes Mgmt. Ass'n, 505 U.S. 88, 98 (1992).
[333] Morales v. TWA, 504 U.S. 374, 383 (1992).
[334] Metropolitan Life Ins. Co. v. Massachusetts, 471 U.S. 724, 739 (1985).

[335] Fidelity Fed. Sav. & Loan Assn. v. De le Cuesta, 458 U.S. 141, 152-53 (1982) (quoting Rice v. Sante Fe Elevator Corp., 331 U.S. 218, 230 (1947)).

[336] Hines v. Davidowitz, 312 U.S. 52, 67 (1941).

[337] S. 2105, §111(a).

[338] S. 2105, §111(b).

[339] N.J. STAT. ANN. §13:1K-19.

[340] MD. ENV. CODE §7-701.

[341] N.Y. EXEC. LAW §713 (2011).

[342] Cf. 15 U.S.C. §7707(b). In the Controlling the Assault of Non-Solicited Pornography and Marketing Act of 2003 (CAN-SPAM), P.L. 108-187, 117 Stat. 2699, Congress included a broad clause preempting all state laws from regulating the use of electronic mail to send commercial messages, but also carved out an exception for any state statute that prohibited "falsity or deception." With no such exception in S. 2105, it may be inferred that Congress intended to preempt the whole field of regulating critical infrastructures.

[343] Chemical Facility Anti-Terrorism Standards, 71 Federal Register 78,276, 78,293 (December 28, 2006).

[344] See President Barack Obama, Remarks on Securing Our Nation's Cyber Infrastructure (May 29. 2009) ("[I]t's now clear this cyber threat is one of the most serious economic and national security challenges we face as a nation."), available at http://www.whitehouse.gov/the-press-office/remarks-president-securing-our-nations-cyber-infrastructure; cf. Michael Jo, National Security Preemption: The Case of Chemical Safety Regulation, 85 N.Y.U. L. Rev. 2065, 2087 (2010).

[345] NATIONAL CONFERENCE OF STATE LEGISLATURES, ELECTRONIC SURVEILLANCE LAW, http://www.ncsl.org/issuesresearch/telecom/electronic-surveillance-laws.aspx.

[346] S. 2102, §8(b). The use of "inconsistent" could be construed as "conflict" preemption language, intended only to displace those laws that would impede DHS's ability to create the information sharing programs.

[347] H.R. 3523, §2(a).

[348] Id.

[349] S. 2151, §102(f)(1); H..R. 4263, §102(f)(1).

[350] NATIONAL CONFERENCE OF STATE LEGISLATURES, ELECTRONIC SURVEILLANCE LAW, http://www.ncsl.org/issuesresearch/telecom/electronic-surveillance-laws.aspx. For a list of state wiretap laws, see appendices in CRS Report 98- 326, Privacy: An Overview of Federal Statutes Governing Wiretapping and Electronic Eavesdropping, by Gina Stevens and Charles Doyle.

[351] See CRS Report R42475, Data Security Breach Notification Laws, by Gina Stevens 4, n.21.

[352] CAL. CIV. CODE §1798.83.

[353] See National Association of Regulatory Utility Commissioners, Information Sharing Practices in Regulated Critical Infrastructure States (2007), http://www.naruc.org/Publications/NARUC%20CIP%20Information%20FIN.pdf.

[354] IND. CODE §5-14-3-4.

[355] ALA. CODE §36-12-40.

[356] MD. CODE ANN. STATE GOV'T §10-615(2).

[357] ARK. CODE. ANN. §23-2-316.

[358] H.R. 3674, §3; S. 2151, §102(f)(3).

[359] S. 2102, §5(d); S. 2105, §704(d).

INDEX

#

9/11, 51, 79
9/11 Commission, 51, 79

A

abatement, 125
abuse, 62, 86, 93, 110, 111
access, 5, 11, 13, 19, 27, 28, 29, 32, 33, 34, 35, 54, 55, 57, 58, 61, 66, 75, 84, 89, 90, 96, 97, 99, 102, 108, 109, 110, 115, 116, 122, 139
accountability, viii, 13, 45, 82, 98
accounting, 11
acquisitions, 40, 46, 60
AD, 72
Administrative Procedure Act, 85
administrators, 39
advocacy, 35
age, 35, 39
agency actions, 85, 128
Air Force, 130, 138
antitrust, viii, 10, 23, 24, 71, 74, 82, 108, 111, 112, 113, 123
APA, 85, 91, 93, 128, 131
appropriations, 28
Appropriations Act, 36, 58, 65, 77
armed forces, 57
assessment, 94
assets, 66, 67, 69, 70, 74, 84, 92, 93, 108, 127, 130, 131, 132, 138
Attorney General, 19, 65, 73, 104, 107, 120, 137
audits, 94
authentication, 49, 67
authenticity, 67
authorities, 9, 10, 11, 13, 19, 22, 26, 33, 36, 37, 43, 88, 95, 102, 105, 118, 122, 129
authority, 2, 4, 5, 11, 13, 15, 16, 22, 25, 26, 27, 28, 32, 35, 36, 39, 40, 43, 46, 53, 54, 55, 59, 65, 68, 69, 70, 73, 74, 76, 77, 83, 85, 87, 92, 94, 95, 104, 105, 117, 118, 119, 120, 123, 133
automate, 97
avoidance, 24
awareness, 5, 9, 45

B

banks, 4
barriers, 15, 52, 60
basic research, 48, 63
benefits, 78, 112
benign, 72
blogs, 71, 137, 138
border search, 103

browser, 75
businesses, viii, 15, 82, 83, 84, 87, 114, 117

C

candidates, 20, 67
case law, 99
cell phones, 99
challenges, 35, 141
Chamber of Commerce, 128
chat rooms, 39
chemical, 2, 6, 44, 65, 83, 128, 130
chemical industry, 2, 6
Chief Justice, 135
chief technology officer (CTO), 41
children, 39
CIA, 45, 105, 106
citizens, vii, 22, 34, 81, 83
City, 99, 100, 133, 134, 135
civil action, 88, 113
civil liberties, 18, 19, 22, 68, 96, 108, 115, 118, 121
classes, 99
classification, 62, 66, 71, 77, 78, 90
Clayton Act, 23, 52
Cloud computing, 35, 75
Coast Guard, 83, 138
Cold War, 28
collaboration, 3, 8, 9, 23, 24, 46, 52, 96, 111, 112
collusion, 15, 23
commerce, 2, 4, 6, 33
commercial, 15, 26, 30, 49, 53, 56, 64, 68, 75, 93, 99, 130, 141
commercial email, 64
common law, 126
communication, 24, 26, 33, 35, 84, 91, 97, 98, 100, 101, 102, 107, 109, 110, 120, 123, 134
Communications Act, 26, 34, 36, 53, 73, 76, 77
Communications Act of 1934, 26, 53, 76, 77
Communications Decency Act, 39, 73, 76
communities, 42, 51, 64, 118
community, viii, 18, 51, 77, 82, 105, 106, 130, 131, 135, 138
compensation, 11
competition, 23, 24, 60, 95, 111, 112
competitive advantage, 114, 117, 123
competitors, viii, 23, 24, 72, 82, 89, 109, 111, 112, 120
complement, 96
compliance, 12, 14, 15, 19, 23, 44, 45, 46, 76, 88, 94, 104, 107, 121, 128, 129
computer, ix, 4, 25, 31, 32, 33, 38, 39, 40, 44, 48, 51, 52, 55, 58, 59, 60, 62, 63, 66, 68, 75, 82, 97, 100, 101, 102, 103, 106, 107, 108, 109, 132
computer fraud, 31, 51, 62
computer systems, 4, 44, 55, 59, 66
computer use, 100, 108
computing, 17, 20, 35, 37, 46, 59, 75, 110, 120, 123
conditioning, 118
confidentiality, 15, 19, 32, 57, 62, 67, 68, 130
conflict, 124, 141
congress, 30, 67
consensus, 15, 26, 27, 66, 92, 98
Consensus, vii, 1, 3
consent, 98, 99, 100, 109, 110, 117, 119, 124, 134, 140
conspiracy, 33
Constitution, 75, 76, 96, 103, 124, 133
construction, 87, 125, 128
consulting, 20
consumers, 50, 111
Continental, 137
coordination, 2, 5, 6, 7, 9, 12, 29, 42, 43, 55, 61
copyright, 39
Copyright, 39
cost, 45, 47, 50, 63, 84, 114

counterterrorism, 29, 36, 55
Court of Appeals, 98
credentials, 67
crimes, 31, 50, 51, 56, 78, 123
criminal statutes, 20, 71
customers, 62, 75, 88, 111, 117, 123
cyber threat information, viii, 81, 83, 111, 115, 116, 117, 118, 119, 120, 121, 123, 124, 126
cyber vulnerabilities, viii, 43, 82, 83
cyberattack, 14, 70
cybersecurity legislation, vii, 1, 5, 7, 14, 84, 91, 104, 115, 127
cyberspace, 6, 28, 29, 54, 66, 68
diplomacy, 28, 54
direct action, 36
directives, 96, 97
directors, 18
disaster, 22
disclosure, 16, 17, 18, 19, 27, 29, 30, 32, 34, 35, 57, 58, 66, 74, 84, 89, 90, 91, 108, 110, 114, 116, 117, 118, 119, 120, 121, 123, 124, 126, 127, 130, 136
disposition, 66
DOJ, 24, 112
domestic laws, 138
draft, 32, 71, 116
Drug Enforcement Administration, 138

D

damages, 18, 94, 113, 118
danger, 110, 137
data center, 20, 68
data processing, 55
DCI, 45
decision makers, 91
deficiencies, 12, 45, 47
degradation, 91, 93
denial, 97
denial of service attack, 97
Department of Commerce, 6, 71
Department of Defense, 6, 28, 36, 69, 72, 90, 105, 106, 131, 135, 138, 140
Department of Energy, 76, 138
Department of Homeland Security, viii, 3, 4, 8, 42, 62, 70, 71, 72, 74, 77, 78, 82, 83, 90, 119, 129, 131, 136, 138
Department of Justice, 6, 19, 20, 24, 72, 73, 75, 90, 101, 112, 130, 137
Department of the Treasury, 138
Department of Transportation, 6
depth, 134
deregulation, 60
destruction, vii, 67, 81, 84, 91, 108, 127
detection, 5, 12, 13, 132
detection system, 132

E

eavesdropping, 4, 30, 34, 56, 58, 136
economic damage, 93
economic security, vii, 81, 84, 127
ECPA, 4, 10, 34, 35, 58, 75, 76, 109, 110, 111, 113, 115, 123, 136
education, 5, 18, 63, 75
educational programs, 11
E-Government Act, 4, 8, 49, 63, 69, 77
electricity, 2, 6
electromagnetic, 26
electronic communications, viii, 34, 36, 38, 58, 82, 96, 99, 100, 108, 109, 126, 136
Electronic Communications Privacy Act, 4, 10, 34, 74, 75, 103, 109, 132, 135, 136
electronic surveillance, 38, 57, 59, 102
e-mail, 35, 75, 99, 100, 101, 102, 133, 134
emergency, 26, 27, 43, 53, 73, 95, 110, 111
emergency preparedness, 95
employees, 90, 96, 98, 99, 100, 101, 102, 114, 116, 129, 134, 139
employment, 85, 100, 101
energy, 84, 125

Energy Policy Act of 2005, 25
enforcement, 6, 14, 31, 35, 38, 43, 45, 56, 62, 68, 85, 112, 120, 128
environment, 25, 36, 66
environments, 3
equipment, 34, 55, 58, 66, 72, 75
espionage, 33
evacuation, 125
evidence, 86, 88, 91, 93, 103, 120, 128, 129
evolution, 3, 72
examinations, 32
exclusion, 115
execution, 124
executive branch, 40, 67
Executive Order, 68
exercise, 86, 87, 88, 95, 105, 129, 137
expertise, 22, 49, 52
exploitation, 15, 17
exposure, 2, 6, 20, 84

F

Facebook, 75, 100
faculty development, 48
fairness, 91
faith, 110, 118, 121
FBI, 22, 129
federal advisory, 31, 56
federal agency, viii, 82, 89, 96, 97, 98, 99, 100, 101, 107, 117, 123, 134
Federal Bureau of Investigation, 131, 138
Federal Communications Commission, 26, 53, 109
federal courts, 86, 98
federal government, vii, ix, 10, 14, 18, 22, 27, 29, 33, 44, 52, 54, 73, 74, 81, 82, 83, 96, 108, 114, 117, 118, 130, 131, 139, 140
federal hiring, 11
federal law, vii, viii, 15, 21, 81, 83, 124, 136
Federal Register, 68, 89, 141
Federal Trade Commission Act, 8, 15, 23
financial, 2, 6, 13, 30, 33, 56, 62, 64, 76, 88, 93, 134
financial incentives, 88
financial institutions, 33, 62, 64
First Amendment, 39, 93
first responders, 44
FISA, 57, 65, 110, 136
flexibility, 87
force, 2, 36, 50, 51, 54, 71
foreign intelligence, 57, 65, 129
foreign investment, 54
foreign policy, 30
Fourth Amendment, viii, 82, 96, 98, 99, 100, 101, 102, 103, 107, 108, 132, 133, 134
France, 73
fraud, 33, 41, 50, 51, 78, 109
freedom, 129, 138
Freedom of Information Laws, 129
funding, 11
funding authorization, 12
funds, 28, 66, 79

G

GAO, 75, 77, 78, 127
General Services Administration, 77
General Services Administration (GSA), 77
globalization, 40
goods and services, 117
governance, 16, 26
governments, viii, 82, 105, 122, 140
grant programs, 30, 56
grants, 31, 48, 63, 70
grid technology, 26
growing frequency, vii, 1
growth, 98
GSA, 55, 60
guidance, 48
guidelines, 6, 10, 18, 49, 55, 59, 64, 105, 107, 108, 112, 116, 118

H

harassment, 39
harbors, 18, 74
hazardous materials, 125
hazardous substances, 125
health, viii, 61, 66
Health and Human Services, 61
health information, 61, 66
history, 87, 128
HIV, 29
HIV/AIDS, 29
homeland security, 4, 51, 64, 125
Homeland Security Act, 8, 42, 73, 76, 77, 95, 97, 130
horses, 97
host, 83
hostilities, 57
House, 2, 3, 5, 7, 8, 9, 11, 19, 25, 26, 32, 33, 37, 43, 46, 47, 67, 68, 70, 71, 72, 73, 74, 75, 76, 78, 79, 92, 106, 116, 119, 129, 138
House of Representatives, 71
hub, 108
human, 129
human resources, 129

I

identification, 27, 99
identity, 12, 25, 41, 50, 51, 61, 64, 65, 67, 77, 78
immigration, 78
immunity, 94, 118, 121, 122, 124
imprisonment, 50
improvements, 11
incidence, 31
indecency, 39, 60
Independence, 65
individuals, 41, 50, 98, 99
industries, 14, 40, 73
industry, 38, 54, 59, 73, 89, 129
inefficiency, 94
inferences, 128
information exchange, 16
information sharing, viii, 2, 3, 5, 6, 7, 8, 10, 11, 13, 16, 17, 18, 23, 24, 42, 43, 68, 74, 77, 82, 97, 105, 106, 108, 112, 113, 115, 117, 118, 119, 121, 122, 126, 131, 139, 141
information technology, 3, 5, 7, 33, 40, 60, 66, 70, 93, 96
infrastructure, vii, viii, ix, 22, 26, 34, 41, 42, 43, 62, 70, 75, 81, 82, 83, 84, 87, 88, 89, 90, 92, 93, 94, 95, 105, 108, 113, 121, 124, 125, 127, 130, 131, 135, 141
injuries, 113
injury, 110
inspections, 94
inspectors, 104
institutions, 48
integrity, 67
intelligence, viii, 4, 27, 33, 37, 40, 51, 54, 58, 64, 73, 75, 76, 77, 82, 95, 105, 106, 116, 118, 131, 135, 137, 138
Intelligence Reform and Terrorism Prevention Act, 51, 71
interagency coordination, 37, 59
interest groups, viii, 82, 98
international terrorism, 129
interoperability, 65
intrusions, 12, 96, 97, 105
invasion of privacy, 126
investment, 69
investments, 11
IP address, 97, 98, 132, 133
isolation, 38
issues, 2, 3, 6, 7, 8, 10, 14, 24, 29, 51, 52, 72, 74, 76, 78, 79, 84, 87, 100, 108, 111, 129, 134, 137
iteration, 97

J

joint ventures, 112
journalists, 57
judiciary, 75, 138
juries, 114
jurisdiction, 24, 26, 40
justification, 29

K

kill, 26

L

law enforcement, 17, 22, 30, 31, 32, 33, 34, 35, 38, 41, 52, 56, 57, 58, 59, 74, 75, 90, 102, 103, 104, 107, 110, 115, 120, 129
laws, vii, viii, 1, 2, 3, 5, 6, 8, 12, 20, 21, 23, 24, 30, 36, 38, 50, 52, 71, 78, 82, 89, 90, 108, 111, 112, 113, 114, 115, 117, 123, 124, 125, 126, 127, 129, 139, 141
lead, viii, 16, 67, 82, 87, 93, 109, 113, 122, 139, 140
learning, 48
legal issues, vii, viii, 68, 81, 82, 83, 91, 104, 116
legal protection, 74
legality, 108
legislation, vii, viii, 1, 4, 5, 7, 14, 26, 72, 81, 83, 84, 91, 104, 115, 124, 126, 127, 139
legislative authority, 41
legislative proposals, 2, 6, 8, 10, 12, 20, 46, 70, 72, 84
light, vii, 1, 25, 26, 37, 51, 100, 105
limited liability, 18
litigation, 85, 90, 114
local government, 20, 42, 62, 90, 95, 105, 121, 124, 127, 140
location information, 35

M

magnitude, 108
majority, 119
malware, 97
man, 52, 129, 133
management, 4, 12, 25, 32, 40, 41, 42, 43, 49, 57, 60, 62, 63
manufacturing, 112
Marine Corps, 138
Maritime Transportation Security Act, 83, 130
Maritime Transportation Security Act (MTSA), 83
mark up, 92, 106, 116, 119
market share, 23
Maryland, 98, 99, 102, 125, 127, 133, 134, 136
mass, 93
materials, 28, 41, 57
matter, 10, 85, 133
media, 28, 74
medical, 18, 83
medical care, 83
Medicare, 78
memory, 75
messages, 53, 83, 97, 100, 141
Microsoft, 76
military, 6, 22, 27, 28, 36, 37, 52, 54, 76, 105, 106, 115, 131, 135, 138
minors, 17
mission, 36, 45, 51, 64, 69, 72, 76, 105
missions, 4, 17, 36
misuse, 15
mobile communication, 35
mobile phone, 75
modifications, 21, 42
money laundering, 79
MTSA, 129, 130

N

National Bureau of Standards, 73
National Center for Missing and Exploited Children, 110
National Defense Authorization Act, 77, 90
national emergency, 26
National Research Council, 66, 69
national security, vii, 4, 5, 6, 17, 27, 39, 40, 45, 51, 54, 59, 60, 64, 69, 81, 84, 93, 95, 105, 106, 116, 117, 118, 119, 123, 126, 127, 139, 141
National Security Agency, viii, 5, 6, 69, 82, 96, 115, 131, 138, 140
National Security Council, 27, 54
national security letters, 139
national strategy, 2, 6
National Strategy, 68
networking, 3, 7
nonfederal systems, vii, 1, 3
nonprofit organizations, 31
North America, 83
NSA, viii, 5, 6, 11, 45, 69, 82, 96, 97, 115, 118, 138, 139

O

Obama, 2, 5, 41, 48, 68, 75, 77, 141
Obama Administration, 2, 5, 41, 48, 68, 75, 77
obstacles, viii, 82, 114, 117, 119
Occupational Safety and Health Act, 129
Office of Management and Budget, 4, 40, 44, 67, 69, 97, 107, 132
officials, 22, 35, 44, 45, 90, 98, 104, 109, 133
OMB, 4, 6, 13, 40, 41, 44, 45, 46, 47, 49, 59, 60, 63, 69, 76, 77, 78, 107, 108
omission, 124, 137
operations, 6, 28, 35, 36, 45, 54, 58, 91, 129
opportunities, 23
organ, 130
outreach, 69
overlap, 67, 95, 105
oversight, 38, 63, 104, 115, 118

P

pagers, 100
participants, 24
password, 33
penalties, 6, 14, 33, 41, 42, 50, 57, 61, 62, 64, 65, 87, 88, 94, 111, 113, 128
permission, 98
permit, 10, 15, 16, 17, 18, 19, 26, 29, 33, 89, 101, 102, 103, 105, 106, 115, 118, 120
Philadelphia, 129
plants, 125
police, 100, 103, 134
policy, 13, 32, 45, 47, 48, 56, 67, 68, 100, 107, 127, 130
policymakers, 108
portfolio, 41
portfolio management, 41
Posse Comitatus Act, 22, 52, 72, 138
postal service, 133
potential benefits, 32
power plants, 83
precedent, 53
predicate, 31, 50, 51, 79
preparation, 11
president, 141
President, 22, 26, 27, 36, 43, 51, 53, 54, 64, 69, 70, 76, 78, 79, 95, 117, 131, 138, 141
presidential authority, 57
prevention, 5, 64, 123, 140
prima facie, 88
primacy, 138
principles, 10, 23, 35, 37, 48, 105, 134
private information, 89

private sector, viii, 3, 5, 8, 14, 29, 40, 43, 44, 49, 52, 74, 81, 82, 83, 88, 95, 105, 108, 109, 113, 114, 115, 116, 117, 118, 119, 121, 123, 124, 126
procurement, 77
profit, 48
propaganda, 28, 39, 54
propane, 128
protected areas, 134
protection, viii, 2, 3, 6, 8, 9, 10, 14, 15, 17, 18, 19, 27, 29, 36, 38, 42, 43, 45, 49, 54, 62, 64, 70, 74, 82, 90, 92, 99, 101, 107, 109, 110, 114, 117, 118, 120, 121, 122, 131, 134
public domain, 90
public health, viii, 81, 84, 127
public interest, 90, 127
public sector, 90, 108, 116, 123, 131
public service, 125
publishing, 39

Q

questioning, 71

R

racketeering, 31, 56
radio, 53
real time, 35, 97
reasoning, 99, 102
recommendations, 2, 7, 19, 21, 50, 51, 72
recovery, 42
recovery plan, 42
reform, 2, 3, 6, 7, 9, 35, 41, 51
Reform, 8, 10, 11, 12, 40, 47, 60, 64, 75, 76, 77, 78
regulations, 14, 15, 61, 62, 64, 65, 70, 84, 127, 129, 133
regulatory agencies, 15, 84
regulatory framework, viii, 2, 81, 84, 127
relevance, 20, 67, 71
reliability, 25, 37, 65, 83
relief, 128
rendition, 109, 110
requirements, viii, ix, 4, 12, 13, 14, 17, 18, 20, 32, 38, 40, 43, 45, 46, 49, 57, 59, 62, 64, 66, 69, 70, 76, 82, 84, 88, 90, 94, 95, 98, 106, 107, 120, 121, 122, 125, 126, 127, 131
RES, 67
resolution, 28
resources, 31, 41, 46, 66, 70, 75, 128
response, 5, 11, 26, 29, 42, 43, 95, 105, 106
restitution, 50, 51, 65
restrictions, 10, 18, 28, 67, 110, 111, 117, 119, 120, 121, 122, 123
RICO, 10, 31, 56, 74
rights, 15, 34, 58, 93, 99, 101, 109, 110, 111, 113, 117
risk, viii, 10, 11, 15, 23, 40, 43, 44, 45, 46, 63, 65, 82, 88, 92, 94, 95, 105, 107, 113, 125, 137
risk assessment, 10, 11, 15, 45, 94, 95, 105
risk management, 44, 63, 125
risks, 5, 14, 15, 20, 37, 52, 70, 92, 94, 112, 113, 114, 131
rules, viii, 35, 47, 82, 91, 96, 98, 100, 114, 125, 129, 138

S

safety, viii, 37, 81, 84, 127, 129
sanctions, 19
Sarbanes-Oxley Act, 62
savings, 125
scholarship, 62
scope, 33, 58, 75, 84, 89, 100, 101, 125, 129, 139
Secret Service, 68
Secretary of Commerce, 4, 13, 40, 45, 46, 47, 59, 60, 63, 69, 76, 77

Secretary of Defense, 19, 27, 36, 45, 54, 121
Secretary of Homeland Security, 11, 12, 13, 14, 15, 16, 18, 19, 32, 40, 42, 46, 65, 70, 86, 92, 94, 95, 118, 121
securities, 51
security practices, 125
seizure, 98, 133
self-expression, 99
Senate, 3, 7, 10, 27, 42, 46, 67
senses, 97
September 11, 51, 124
service provider, 17, 35, 39, 60, 98, 109, 110, 111, 133
service quality, 109, 111
services, 4, 17, 27, 35, 39, 40, 49, 60, 63, 70, 75, 93, 107, 109, 110, 121, 123, 126
SFS, 69
Sherman Act, 111
showing, 35
sobriety, 103
Social Security, 78
society, 98
software, 40, 48, 66, 69, 113
specifications, 48
speech, 15
stakeholders, 35
state, viii, ix, 20, 24, 30, 31, 42, 56, 62, 68, 81, 82, 83, 88, 89, 90, 103, 105, 108, 109, 113, 114, 115, 117, 118, 121, 124, 125, 126, 127, 129, 140, 141
state laws, ix, 82, 108, 113, 125, 126, 141
state legislatures, 103
states, 10, 24, 73, 74, 124, 125, 126, 127
statistics, 74
statutes, vii, 1, 4, 20, 21, 23, 50, 56, 71, 85, 89, 96, 103, 113, 125, 130, 132, 138
statutory authority, 69, 71, 76, 104
storage, 35, 109
strategic planning, 37
stress, 37
structure, 35, 124
subpoena, 35, 36
subscribers, 54
supply chain, 5, 38, 40, 43, 46
Supreme Court, 85, 86, 87, 98, 99, 100, 102, 111, 129, 134
surveillance, 59, 98, 102, 134, 135
survival, 54

T

target, 38, 111
Task Force, 7, 8, 9, 10, 11, 12, 13, 14, 16, 17, 18, 20, 21, 24, 30, 31, 34, 36, 38, 44, 46, 51, 68, 70, 71, 72, 73, 74, 75, 76, 78, 79, 140
technical assistance, 11, 42, 44, 45, 62
techniques, 98
technologies, 34, 38, 42
technology, 3, 5, 26, 35, 44, 48, 66, 93, 105
technology transfer, 44
telecommunications, 34, 38, 39, 58, 59, 60, 66
Telecommunications Act, 108
telecommunications services, 34, 58
telephone, 99, 102, 132, 133, 136
telephone numbers, 102, 133
territorial, 28, 54
terrorism, 42, 47, 63, 83
terrorist acts, 78
terrorist attack, 51
testing, 44, 97
The Homeland Security Act, 4, 11
theft, 2, 6, 41, 50, 51, 61, 64, 77, 78
threats, viii, 12, 15, 17, 22, 23, 24, 26, 27, 32, 43, 45, 52, 73, 79, 81, 82, 83, 84, 88, 92, 95, 96, 97, 105, 108, 111, 112, 113, 119, 120, 123, 131, 136
Title I, 10, 30, 34, 44, 50, 56, 69, 77, 92, 95, 104, 105, 106, 121, 122, 125, 131
Title II, 10, 30, 34, 44, 50, 56, 69, 77, 95, 104, 105, 106, 121, 122, 131

Title V, 39, 73, 77, 119, 121, 122, 126
tracks, 129
trade, 15, 23, 24, 30, 52, 56, 61, 90, 111, 114, 120, 137
trafficking, 33, 58, 75
training, viii, 44, 45, 49, 59, 82, 98, 104
transactions, 68
transmission, 25, 53, 64, 99, 109, 125
transparency, 30, 89, 115, 118
transportation, 6, 84, 93
Treasury, 131, 135
treatment, 134
trial, 86
triggers, 97

U

U.S. economy, 26, 93
uniform, 74
United, vii, 1, 3, 22, 26, 28, 29, 33, 39, 43, 50, 51, 52, 55, 65, 66, 71, 72, 76, 79, 86, 93, 97, 98, 99, 101, 102, 104, 124, 132, 133, 134, 135, 136, 137, 140
United States, vii, 1, 3, 22, 26, 28, 29, 33, 39, 43, 50, 51, 52, 55, 65, 66, 71, 72, 76, 79, 86, 93, 97, 98, 99, 101, 102, 104, 124, 132, 133, 134, 135, 136, 137, 140
unreasonable searches, 96, 98, 133
USA, 21, 62, 136
USA Patriot Act, 21
USA PATRIOT Act, 62, 136

V

variations, 12, 46
vessels, 83, 129
victims, 41, 50, 51, 65
vote, 70, 92, 119
vulnerability, 25, 40, 42, 53, 62, 74, 116, 130

W

waiver, 90, 114, 120
war, 26, 68
Washington, 66, 69, 78, 129
water, 83, 130
water supplies, 83
web, 35, 39, 60, 75
web pages, 75
websites, 39, 60, 103, 133
White House, 5, 6, 7, 8, 9, 10, 11, 12, 13, 14, 15, 16, 17, 19, 20, 21, 30, 31, 33, 40, 41, 43, 46, 47, 49, 66, 67, 68, 70, 73, 74, 77, 78
wholesale, 111
word processing, 35
workforce, 2, 6, 9, 11
workplace, 99
worms, 97
wrongdoing, 103, 134

Y

yield, 37